Studies in Computational Intelligence

Volume 696

Series Editor

Janusz Kacprzyk, Polish Academy of Sciences, Warsaw, Poland

The series "Studies in Computational Intelligence" (SCI) publishes new developments and advances in the various areas of computational intelligence—quickly and with a high quality. The intent is to cover the theory, applications, and design methods of computational intelligence, as embedded in the fields of engineering, computer science, physics and life sciences, as well as the methodologies behind them. The series contains monographs, lecture notes and edited volumes in computational intelligence spanning the areas of neural networks, connectionist systems, genetic algorithms, evolutionary computation, artificial intelligence, cellular automata, self-organizing systems, soft computing, fuzzy systems, and hybrid intelligent systems. Of particular value to both the contributors and the readership are the short publication timeframe and the world-wide distribution, which enable both wide and rapid dissemination of research output.

Indexed by SCOPUS, DBLP, WTI Frankfurt eG, zbMATH, SCImago.
All books published in the series are submitted for consideration in Web of Science.

More information about this series at http://www.springer.com/series/7092

Weldon A. Lodwick · Luiz L. Salles-Neto

Flexible and Generalized Uncertainty Optimization

Theory and Approaches

Second Edition

 Springer

Weldon A. Lodwick
Department of Mathematical
and Statistical Sciences
University of Colorado Denver
Denver, CO, USA

Luiz L. Salles-Neto
Department of Science and Technology
Federal University of São Paulo
São Paulo, Brazil

ISSN 1860-949X   ISSN 1860-9503   (electronic)
Studies in Computational Intelligence
ISBN 978-3-030-61182-8   ISBN 978-3-030-61180-4   (eBook)
https://doi.org/10.1007/978-3-030-61180-4

This Springer imprint is published by the registered company Springer Nature Switzerland AG
The registered company address is: Gewerbestrasse 11, 6330 Cham, Switzerland

Preface

The second edition of *Flexible and Generalized Uncertainty Optimization: Theory and Approaches* is updating of the first edition where an amplified fuzzy interval minimax regret subsection included as well as a new section on generalized uncertainty robust optimization. Typos that have come to our attention associated with the first edition have been correct.

This book presents the theories and methods of flexible and generalized uncertainty optimization. The first chapter contains an overview of flexible and generalized uncertainty optimization. Given that generalized uncertainty is a relatively newer theory compared with fuzzy set theory and given that fuzzy set theory is well-developed with excellent textbooks (see for example [1, 2]), fuzzy set theory is not explicitly reviewed. We present in Chap. 2 an outline of the relevant theory of generalized uncertainty as it pertains to optimization modeling. Chapter 3 has the construction methods requisite for obtaining flexible and generalized uncertainty input data in such a way that the data can be used in flexible and generalized uncertainty optimization model. Chapter 4 has an overview of the common themes of flexible and generalized uncertainty optimization. Chapter 5 focuses on flexible optimization theory. Chapter 6 presents generalized uncertainty optimization.

This book presents the theory and methods of flexible and generalized uncertainty optimization. The generalized uncertainties that we cover are uncertainties associated with lack of information, and as developed here, are more general than stochastic theory where well-defined distributions are assumed. As this term is used in this text, generalized uncertainty entities are families of distributions rather than a single distribution. Our interested are families that are enclosed by upper and lower functions.

Denver, USA

Weldon A. Lodwick

References

1. B. Bede, The Mathematics of Fuzzy Sets and Fuzzy Logic. (Springer, 2012)
2. G. J. Klir and B. Yuan, Fuzzy Sets and Fuzzy Logic: Theory and Applications. (Prentice Hall, New Jersey, 1995)

Contents

Chapter 1
An Introduction to Generalized Uncertainty Optimization

1.1 Introduction

Our primary interest is a unified theory of flexible and generalized uncertainty optimization at the level of a beginning to intermediate graduate student and an experienced optimization practitioner. Thus, it is assumed that the reader is familiar with basic notions about optimization theory. Though it is helpful if the reader has some basic understanding of fuzzy set theory, probability theory, and interval analysis, this text will develop much of what is needed to understand the content of this book.

Given that this textbook's focus is in optimization, we are interested in quantitative methods and will restrict ourselves to real numbers $\mathbb{R}$. Let us start with an example of the flexibility and generalized uncertainty that introduces the ideas and approaches of the monograph.

Suppose we have incomplete information about the supply and price of crude oil coming from the middle east due to a variety of factors such as stability of oil production and civil war. Moreover, suppose we are concerned about producing enough to meet the demand for the various grades of gasoline in central Germany for the month of January. How do we model an input of "There will be more or less 500,000 barrels of Middle Eastern crude oil available at more or less x euros per barrel"? How does one mathematically analyze, "The market demand in central Germany for high grade gasoline is about 400,000 L in January"? How does one interpret to a production manager, "We need to produce about 300,000 L of regular gasoline to sell in the month of January"? Lastly, suppose the CEO says, "The company needs to exceed or come as close as possible to the income we obtained in our best profit year." How does one mathematically encode "more or less", "about", or "exceed or come as close as possible" in a mathematical model if no more precision of the data or information is given?

This book will look at how to be mathematically precise in the midst of these types of imprecisions, fuzziness, and uncertainty in the information composing such a problem:

© Springer Nature Switzerland AG 2021
W. A. Lodwick and L. L. Salles-Neto, *Flexible and Generalized Uncertainty Optimization*, Studies in Computational Intelligence 696, https://doi.org/10.1007/978-3-030-61180-4_1

- In the inputs (amount of crude oil available);
- In the processing (number of liters produced of various grades);
- In the objective (profit to be made by the production).

At each phase (input, production, objective) there are a variety of uncertainties many or most of which are *not probabilistic*! For example, the supply of oil from Iraq in the midst of internal civil war affects the number of barrels and price of middle eastern crude oil in a non-probabilistic way. That one can and analysts do use probability is clear.

However, this monograph is devoted to the proposition that the most useful way to model inherently non-probabilistic inputs and outputs is to mathematically analyze the problem from the beginning in its "native" environment of non-probabilistic uncertainty, which, in many cases, involves fuzzy sets, possibility/necessity pairs, interval-valued probability, or intervals. The latter three types are what we will define as *generalized uncertainty*. To be sure, in real models, a decision needs to be made. For example, in the above, it is clear that production of gasoline for central Germany will occur, that is, gasoline will be supplied in a particular quantity in central Germany in the month of January from zero production to over-production. Our philosophy, like that of recourse models in stochastic optimization, is to carry all the uncertainty (or as much as possible) *in its "native" form* until a deterministic decision needs to be made (go or no go, 150,000 L or 160,000 L, and so on).

Our objective in this first chapter is to classify newer types of uncertainty optimization into two distinct types of optimization, flexible optimization, and optimization under generalized uncertainty. One type of flexible optimization is fuzzy optimization. One type of optimization under generalized uncertainty is possibilistic optimization. These two types of optimization—fuzzy, possibilistic—are the two generic types most often found in books and journals. We will postpone a more detailed discussion about fuzzy set theory, possibility theory, and theory of generalized uncertainty to Chap. 2 and the full optimization development to Chaps. 4–6. Here, we give a brief introduction to fuzzy set theory and possibility theory to help us understand the efficacy of flexible and generalized uncertainty optimization. The ensuing sections of this first chapter will introduce the following ideas, which will be developed in more detail in subsequent chapters:

1. What is meant by flexibility, generalized uncertainty and the power and usefulness of these in optimization (Sect. 1.2);
2. The mathematical language of flexibility being fuzzy set theory and the mathematical languages of generalized uncertainty being possibility theory, interval-valued probability, or probability-based possibility (Sect. 1.3);
3. The taxonomy for flexible optimization and optimization under generalized uncertainty which leads to distinct theoretical and algorithmic methods (Sect. 1.4).

Definition 1 A **fuzzy set** is defined by its membership function which, for this monograph, is restricted to be a single-valued continuous concave function

$$\mu : X \subseteq \mathbb{R} \to [0, 1]. \tag{1.1}$$

Remark 2 The condition that a membership function be continuous can be relaxed. For this monograph, this generalization is not used. Fuzzy sets are sets for which belonging is gradual, not Boolean (classical), and this gradualness is indicated by the range of the membership function being $[0, 1]$. A classical set, described via a membership function, has a range $\{0, 1\}$, either an element belongs to the set or it does not.

Remark 3 One of the concerns that has been expressed by some researchers with respect to fuzzy sets is its mathematical meaning, that is, the mathematical semantics of fuzzy sets. Given the flexibility and often ad hoc way that fuzzy sets are presented via its membership function, a membership function which can be anything a user desires it to be, the concern is that the membership function is purely a heuristic entity and therefore outside classical mathematics. For this monograph, as noted above, a fuzzy set, defined by its membership (1.1), is a generalization of the definition of set. Instead of a set being defined via a Boolean set belonging function, the characteristic function,

$$\mu_A(x) \equiv \chi_A(x) = \begin{cases} 1 & x \in A \\ 0 & x \notin A \end{cases}, \quad A \text{ a classical set,} \tag{1.2}$$

a fuzzy set is given a gradual set belonging defined by (1.1). Clearly, fuzzy sets are a generalization of classical mathematical sets and its semantics is that associated with sets albeit gradual sets. One might level the same criticism to the way a classical set is defined as "a collection of objects" as being "ad hoc" and "heuristic" since a set can be anything a user desires it to be. Nevertheless, it is considered as well-defined and the basis of much of mathematics. So, to reiterate, for this monograph a fuzzy set is a generalization of set theory and its mathematical semantics is that of generalized set theory. In the *applications* of fuzzy sets to problems, the way gradualness is chosen and implemented may be heuristic. However, as a mathematical object, a fuzzy set is a set theoretic object and so its semantic is well-defined in the sense of generalized set theory.

Notation 4 A word about our notation of fuzzy and generalized entities. The notation for a fuzzy entity A is a tilde, $\tilde{\ }$, over it, $\tilde{A}$, and a generalized uncertainty entity B is a circumflex, $\hat{\ }$, over it, $\hat{B}$. A generic fuzzy set $\tilde{A}$ is denoted here as its membership function $\mu(x)$. Often, we write a subscript, $\mu_{\tilde{A}}(x)$, to indicate that the fuzzy membership function is of a particular fuzzy set, $\tilde{A}$ though the tilde is usually left of the membership function so the membership function is denoted $\mu_A(x)$. When there is no designation as to whether a set is a fuzzy set or an uncertainty entity, it will refer to either where the context will determine its semantic.

Flexibility, as we will use it here, has as its focus *constraint set belonging*. From our point of view, all that we call flexible optimization arises from the relaxation of constraint set belonging, the relationship "belongs to", $\in$. This is precisely what fuzzy sets encode (see [1]). Thus, flexible optimization as developed here includes what has been traditionally called fuzzy optimization.

Flexible optimization and optimization under generalized uncertainty are relatively newer approaches to optimization developed over the last four decades. They are worthy of serious consideration in that the very nature of many optimization models inherently have underlying flexibility and generalized uncertainty. The first researchers to distinguish fuzzy (flexible) from possibilistic (a type of generalized uncertainty) optimization and to use the term "*flexible optimization*" in a publication were Inuiguchi and Tanaka [2] who use this term to describe "optimization problems with vagueness." However, the first researcher, to our knowledge, to use the term "flexibility" in relation to fuzzy optimization was Hans J. Zimmermann (personal communication with Professor M. Inuiguchi).

Generalized uncertainty, as the term is used in this text, is limited to uncertainty that arises from partial information. Generalized uncertainty of interest, as we shall see, can be represented by a set of distributions bounded by an upper and lower pair of distributions or a single distribution. Uncertainty, in these cases, exists because we do not know which distribution from the set of all distributions in the set bracketed by the bounding functions is the appropriate one.

We begin by formalizing what we mean by flexibility and generalized uncertainty.

Definition 5 Flexibility in the context of an optimization problem, is the relaxation of the meaning of set relationships such as belonging (to the constraint set) or optimizing (in the objective). That is, the relationships "belongs to", $\in, =, \leq, \min,$ or $\max$ take on a gradual rather than a Boolean or absolute meaning.

When belonging takes on the meaning of "come as close as possible but do not exceed" then there is flexibility. For example, one may have a deterministic constraint of a fixed upper bound on the hours of labor based on the number of employees a company has available. However, it is often possible and no doubt wise to have a pool of overtime and/or temporary laborers that the company can draw on at a perhaps increased cost.

Fuzzy sets have a semantic of gradual, non-Boolean, set belonging so that it is a natural mathematical language for problems in which flexibility is an inherent part of the modeling and analysis. Flexibility includes fuzzy sets but in the sense we are using the word, it includes things such as "come as close but do not exceed". These types of flexibilities are not fuzzy sets *per se*, but can be translated into fuzzy sets.

Definition 6 Generalized uncertainty theory is a mathematical theory of incompleteness or lack of information, lack of specificity, or imprecision.

The above definition of generalized uncertainty used the word "uncertainty" and "incompleteness of information" in its definition. These words for this monograph will have a restricted meaning that we articulate next.

Definition 7 ([3]) **Uncertainty** is the state of not knowing the exact value of an entity or the truth of a proposition/statement. That is, a piece of information or data is said to be uncertain for an agent when the agent does not know whether or not the piece of information is true or false.

Example 8 In a computer program, suppose we have the following instructions: IF $x > 0$ THEN $y = 3$, IF $x < 0$, THEN $y = -3$, OTHERWISE $y = 0$. When the computer branches and returns $y = 0$, do we know it was caused by a real-value 0? Since the computer is subject to round off error, there is an uncertainty associated with the branching.

Definition 9 ([3]) A piece of information or data is said to be **incomplete** (or imprecise, not completely specified, lacks information) in a given context if it is not sufficient to allow the agent to answer a relevant question in this context.

Example 10 Let set $A \equiv Senior\_Aged = \{65, ..., 90, ..., 120\}$. When we are asking for the age of a person X and all we know is the fact that $X \in A$, this piece of information is an incomplete piece of information, since we are unable to answer the question about the age of person X. However, if the question were, "Is this person X over 59?" then the information $X \in A$ is not incomplete, that is, complete.

The generalized uncertainties that are of interest to this text are only those that can be represented by bounding *pairs* of set-valued functions or distributions and single uncertainty distribution functions. Single functions are either derived as a model of the uncertainty itself, or chosen as an approximation between the bounding pair. Whereas the mathematical language of flexibility used in this text is fuzzy set theory, the mathematical languages of generalized uncertainty are many and include:

1. Generalized uncertainties directly translating into pairs of functions that bound the uncertainty

    (a) Intervals [4, 5];
    (b) Fuzzy Intervals [6, 7];
    (c) Possibility/Necessity Measures [8];
    (d) Interval-Probability [9];
    (e) P-Boxes [10, 11];
    (f) Clouds [12, 13];

2. Generalized uncertainty that bounds the uncertainty with a confidence limit— Kolmogorov, Smirnov Statistics [14];
3. Generalized uncertainties that require further hypotheses to translate into pairs of distributions that bound the uncertainty

    (a) Belief/Plausibility Measures [15, 16];
    (b) Probability Intervals [17];
    (c) Random Sets [18].

Remark 11 Imprecise Probability [19] is type of generalized uncertain which is an alternative to Belief/Plausibility However, we do discuss this since its development is beyond the focus of this presentation and we will concentrate of the use of Interval Probability as the unifying theory for generalized uncertainty.

Each of these representations of incomplete or partial information is or can be represented by a pair of bounding functions under some conditions that will be specified. Note that given a pair of bounding functions, a single function can be chosen as an approximation of the correct uncertainty along with an associated error just as we can choose the midpoint as approximating a value in an interval where the maximum error of choosing the midpoint is half the width of the interval. We focus our presentation on the six items of (1) and the one item of (2) above since the three items of (3) can be transformed into (1d), although, in reality, the six items of (1) can be transformed into item (1d), interval-valued probability. We show how to translate a few of the items in (3) to demonstrate how to translate the data into interval-valued probabilities given additional hypotheses on the data. However, we will assume that the data will have sufficient structure to be in the first two classes above.

What is the salient feature of generalized uncertainty is that it can be characterized by two bounding functions that contain a set of distributions that describe the uncertainty, the lack of information. Stochastic problems, on the other hand, are characterized by a single distribution function. Generalized uncertainties are characterized by a family of distributions enclosed by two bounding functions or distributions. In some cases a single distribution that describes the uncertainty, which may be used to generate a pair of bounding distributions. Equations (1.3) and (1.4) of Example 13 below, illustrate a pair of bounding distributions. These bounding functions, in turn, can also generate a single distribution that encodes the uncertainty as can be seen in Eq. (1.5).

On the other hand, given Eq. (1.5), a single distribution encoding incomplete information, Eqs. (1.3) and (1.4) can be constructed. The key feature of generalized uncertainty are the pair of enclosing distributions. When the upper and lower distributions are equal, then we have the stochastic case when the bounding functions are probability distribution in precisely the way it happens with intervals when the right interval and the left interval are equal changes interval analysis into real analysis.

The class of distributions comprising the first six generalized uncertainty of the first class are characterized by two bounding distributions that guarantee the enclosure of the uncertainty. The two bounding distributions generated by the distribution in the second class are guaranteed to enclose within a *confidence limit*. In addition, in many cases, given a bounding set of distributions, one can generate a corresponding fuzzy interval.

Given that we have associated, with uncertainty data, an underlying associated fuzzy interval, there are *three* possible ways we can use generalized uncertainty input data in an optimization problem:

1. The fuzzy interval itself that defines the uncertainty;
2. The bounding pair (upper or lower) of distributions or any approximation within the bounding distributions;
3. Both bounding functions simultaneously.

The use of the first or second of these in an optimization will result in flexible optimization in the case of a fuzzy semantic or possibilistic (optimistic or pessimistic or approximate) optimization that is like a generalized recourse type of optimization.

The third type only pertains to generalized uncertainty optimization and results in a minimax, penalized, or robust type of optimization.

Let us look at examples of fuzzy and generalized uncertainty data.

Example 12 (*Fuzzy data*) The set of voxels that delineate a tumor on a computed tomography (CT) image is transitional (at least for some voxels) since there are voxels that are in transition from non-tumor to tumor. That is, on the universal set of the voxels of a CT image, the set of voxels "tumor" is a set characterized by transition from non-tumor to tumor. Some voxels may be both tumor and non-tumor to a given degree simultaneously. That is, they are simultaneously tumor and non-tumor.

Example 13 (*Uncertainty data*) Suppose we know that the cumulative probability of the cost of a product is between two cumulative distribution functions $\underline{C}, \overline{C}$ where

$$\underline{C}(x) = \begin{cases} 0 \text{ if } x < 5 \\ 1 \text{ if } x \geq 5 \end{cases} \tag{1.3}$$

and

$$\overline{C}(x) = \begin{cases} 0 \text{ if } x < 7 \\ 1 \text{ if } x \geq 7 \end{cases} . \tag{1.4}$$

Suppose this is the only information we have about the cost, that is, the cost is between $5 and $7. The actual cost, represented as a cumulative distribution, is a cumulative distribution between $\underline{C}$ and $\overline{C}$ but we have no idea what it is given that this is the only information we have. Indeed

$$CDF(x) = \{C(x)|\underline{C}(x) \leq C(x) \leq \overline{C}(x)\}$$

is uncertainty data.

Note that in the first example, the *elements* (voxels) are clearly defined, but the set is fuzzy, gradual and inherently so. In the second example, the set in question, cost of the product we will pay, is clearly defined but the element, the particular cumulative distribution which is the cost, that is, the one which is the cost we will actually pay, is not clear. It is clear that no business charges the interval [$5, $7]. A business charges a single real number payment between $5 and $7. The only evidence available for this example about the actual cost is not a real number, deterministic, but an interval. The *set* [5, 7] is precisely defined/known but the element (the actual charge) is not.

Another characteristic that can be understood by these two examples is that for a genuine fuzzy set such as *tumor*, it is not a matter of resolution. The boundary of tumor and non-tumor remains gradual regardless of the resolution. The gradualness of a tumor will always remain. That is, the fuzziness remains even with more data. However, if we have more information about the cost, it will shrink the interval [5, 7], and we will have more precise information about the actual cost. More information in generalized uncertainty is monotonic (isotonic in the language of interval analysis).

- Fuzzy sets are represented by a **unique** membership function [1] whereas partial information as encoded by a **pair** of bounding distributions or measures enclosing the set of all unknown distributions or measures generated by the incomplete information such as Eqs. (1.3) and (1.4) or a single distribution that encodes the lack of information. For example, [5, 7] is an encoding of lack of information and as an entity describing this lack, it can be considered as

$$f(x) = \begin{cases} 0 \text{ if } x < 5 \text{ or } x > 7 \\ 1 \text{ if } 5 \leq x \leq 7 \end{cases}. \tag{1.5}$$

On the other hand, Eqs. (1.3) and (1.4) represent a pair of functions that are bounds on the uncertainty represented by the interval [5, 7].

- Fuzzy sets are sets that model gradual set belonging described as a single-valued function*. Uncertainty is a state in which the information at hand does not allow the determination of the truth or value as a unique real number of the entity under discussion.
- Variability and frequency are modeled by probability distributions which require precise information of all possible states represented by a single real-valued distribution function, or its probability density function.
- If we are dealing with upper and lower distributions that are cumulative distribution functions (CDFs), probabilities can be thought of as "degenerate" distribution pairs where the upper bounding function is equal to the lower bounding function, just as a real number can be thought of as a degenerate interval in which the upper bound is equal to the lower bound of the interval.
- Each of the generalized uncertainty types of interest to this exposition has the property that with more precise information about the uncertainty, the closer the pair of distributions approximate a degenerate single distribution.

* There are type-2 fuzzy sets (and other types) which are not single-valued functions. Type-2 fuzzy sets are described by a set of single-valued membership functions, one for each element of the range (every element of the range is itself a fuzzy set). However, this monograph does not study these types of fuzzy sets but the interested reader can consult [20].

Optimization Problems: A real-valued (deterministic) optimization model is a *normative* mathematical model whose underlying system is most often *constrained* and its general form is:

$$z = \text{opt } f(x, c) \tag{1.6}$$
$$\text{subject to } g_i(x, a) \leq b_i \ i = 1, ..., M_1, \tag{1.7}$$
$$h_j(x, d) = e_j \ j = 1, ..., M_2, \tag{1.8}$$
$$x \in S \subseteq \mathbb{R}^n. \tag{1.9}$$

We denote the constraint set by

$$\Omega = \{x | g_i (x, a) \leq b_i \ i = 1, ..., M_1, h_j (x, d) = e_j \ j = 1, ..., M_2,$$
$$x \in S \subseteq \mathbb{R}^n\}.$$

It is assumed that $\Omega \neq \emptyset$. The values and/or nature of of $a, b, c, d, e, \leq, =, \in$, and opt (min / max, other such as "come as close as possible") are the input (data) for the optimization model. Our general model can be symbolically reformulated as

$$z = \text{opt } f(x, c) \tag{1.10}$$
$$x \in \Omega (a, b, d, e, S), \tag{1.11}$$

where we denote the constraint set as a function of the input parameters for emphasis.

Equality is equivalent to "greater or equal to" *and* "less than or equal to" for real variables. So there is no mathematical need for listing these (1.8) separately. However, this is not true for intervals and hence for fuzzy intervals since fuzzy intervals are described as an uncountable set of intervals called α−cuts. Moreover, intervals are a type of fuzzy set and at the same time may be considered as generalized uncertainty depending on its semantics. Already, the reader is alerted to the fact that we are dealing with entities that have a different structure than that of real numbers as the following example illustrates.

Example 14 Suppose we have the interval equation,

$$[1, 2]x = [2, 4],$$

and we wish to solve it by two inequalities, $\leq$ and $\geq$. For

$$[1, 2]x \leq [2, 4],$$

the solution, using standard/classical interval analysis, is,

$$x \in (-\infty, 1].$$

For

$$[1, 2]x \geq [2, 4],$$

the solution, using interval analysis is,

$$x \in [4, \infty).$$

Thus,

$$[1, 2]x \leq [2, 4] \ and \ [1, 2]x \geq [2, 4]$$

yields

$$x \in (-\infty, 1] \cap x \in [4, \infty) = \emptyset.$$

On the other hand, it is clear that using classical interval analysis,

$$[1, 2]x = [2, 4] \Rightarrow x = [2, 2].$$

Thus, we distinguish between equality and inequality in the representation of mathematical programming problems whose constraint sets are made of flexibility or generalized uncertainty coefficients.

We have noted that when the $\in$, "element of Ω", is relaxed, we will have flexible optimization. Note that from this point of view (we will make this precise in the next chapter), all that has been called "soft constraints," including the meaning of "optimize" as it relates to the objective function or optimization with vagueness, can be considered as a relaxation of set belonging. If the objective function is a soft constraint, it is a target objective which is given or calculated. This target is then transformed into a soft constraint and a deterministic objective function is used to minimize variations from the target. The precise way to do this will be given in Chap. 4. When some or all the values of a, b, c, d and e of (1.10) and/or (1.11) are generalized uncertainty types, we have optimization under generalized uncertainty. There are generalized notions of convexity (see, for example, [21]) which we do not pursue in this text. However, we note that the usual solution methods are often local in optimization under generalized uncertainty even if the deterministic equivalent model is convex since a generalized uncertainty constraint set equivalent of a deterministic convex constraint set is generally not convex. For example, in a very simple case where the constraints are linear of the form $Ax \le b$, and the coefficients of the matrix are intervals, that is, if we are given

$$[2, 4]x_1 + [-2, 1]x_2 = [-2, 2]$$
$$[-1, 2]x_1 + [2, 4]x_2 = [-2, 2],$$

then, the solution set can be a star-shaped region of Fig. 1.1 (see [22]) which is not convex.

There is another point of view in which the optimization problem statement begins as a fuzzy optimization statement rather than a real-valued optimization model some or all of whose parameters are fuzzy and/or intervals and/or possibility (see [23, 24]). For the purposes of this text, we begin with a model that exists as a real-valued model (1.10), (1.11) but whose inputs are flexible or generalized uncertainty types. In the case of generalized uncertainty, we assume the problem converges to (1.6)–(1.8) when the parameters converge to real numbers, that is, when the uncertainty is eliminated.

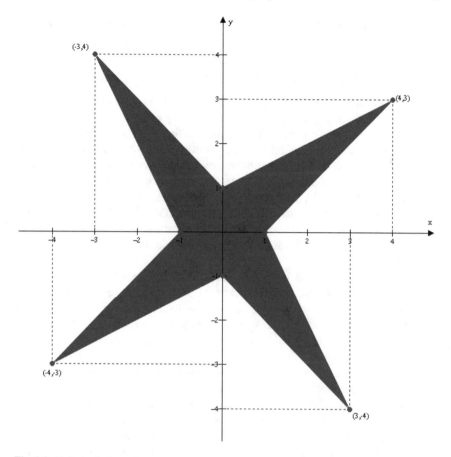

Fig. 1.1 United solution set

1.2   What Interval, Fuzzy, and Generalized Uncertainty Optimization Contribute to Optimization Theory

We make a case for flexible (fuzzy) optimization and optimization under generalized uncertainty as a separate study by looking at the basics of optimization modeling. Not only this, but it is our thesis that flexible and generalized uncertainty optimization play a central role in more realistic normative problems compared to deterministic ones.

1.2.1 The Importance of More General Types of Optimization

This text considers the following three principles crucial to keep in mind in the process of optimization modeling. Our three principles are the following.

1. Optimization models of real systems are very often *satisficing* (see [25, 26]) in which case fuzzy and generalized uncertainty methods are key approaches to these optimization models. Satisficing as defined by Herbert Simon (see [25, 26]), means that decision makers rarely work with the deterministically "best" solution or are even able to obtain "the best" solution to a real problem, but seek to obtain solutions that are satisfying. Solutions that are satisficing are inherently flexible. Clearly, the usual deterministic models, if used to model decision processes described by Simon, need to be modified. Flexible and generalized optimization, as we shall see, are able to model satisficing in a natural and direct way. They are mathematical languages that can and do encode satisficing problems.

2. Many optimization models are *epistemic*. That is, models that are epistemic are those which we, as humans, construct from knowledge about a system rather than models that are constructed from the system itself. For example, an automatic pilot of an airplane models the system physics. A fuzzy logic chip that controls a rice cooker is a model of what we know about cooking rice rather than the physics of rice cooking.

3. Very often input data that compose the parameters of the model arise from incomplete information whose completion may be theoretically possible but practically impossible. Some data, such as projections (for example, the future demands for a product or what will happen to supplies in the midst of civil unrest) are inherently incomplete and uncertain. *Possibility* theory (see [27]) is one mathematical language that encodes partial information and is one type of what we call generalized uncertainty.

That deterministic or stochastic approaches can be and are used in models, which have satisficing, epistemic, and incomplete information, is not disputed. Stochastic models, in the final analysis, will translate probabilities into a deterministic system. Our methods will translate into deterministic systems as well. Our point is that the formulation of the model, from the beginning, needs to be as close as possible to its natural setting (for example satisficing, epistemic, or possibilistic) and these representations need to be carried as long as possible rather than translating the problem directly into deterministic or stochastic models right at the start.

Next, a simple example illustrating the difference between taking the average from the start and carrying the uncertainty as long as possible is presented. Other examples can be found in Jamison [28, 29].

Example 15 Consider the problem of evaluating the following expression,

$$[1, 2]x + [3, 4]x = [4, 12], x \geq 0.$$

Choosing the midpoints (averages),

$$\frac{3}{2}x + \frac{7}{2}x = 8, x = \frac{8}{5}.$$

Carrying the uncertainty,

$$[4, 6]x = [4, 12],$$
$$x = [1, 2]$$
$$x_{midpoint} = \frac{3}{2}.$$

First the solution $x = [1, 2]$ is clearly more informative with respect to the in the uncertainty of the solution than $x = \frac{8}{5}$. Secondly, the results are different with the $\frac{8}{5}$ average weighted more toward the right endpoint of the resulting uncertainty and so over estimating the $\frac{3}{2}$ average. See [28, 29] for further discussion.

One begins any modeling process by stating the problem in its native setting which, in the context of our discussion, means explicit inclusion of flexibility and generalized uncertainty when these are present. Since it is our thesis that many, if not most, problems in optimization are flexible, satisficing, information deficient, and/or epistemic as opposed to fitting all optimization models into a deterministic or strictly stochastic model, the natural mathematical languages in which these types of problems may be stated is that of fuzzy set theory and generalized uncertainty theory. Our point of view is that optimization models that represent satisficing epistemic or lack of information and are normative processes have two natural mathematical languages in which the problem may be stated, fuzzy set theory and generalized uncertainty (possibility) theory.

Rommelfanger [30] (p. 295) states that arguably the most widely used operations research method is linear programming. He goes on to state that even though this is true, of the 167 production (linear) programming systems investigated and surveyed by Fandel (see [31]) only 13 of these were "pure" deterministic linear programs. Thus, Rommelfanger concludes that even with this most highly used and applied operations research method, in most cases there is a discrepancy between classical deterministic linear programming and what is actually done.

Deterministic and stochastic optimization models require well-defined input parameters (coefficients, right-hand side values), relationships (inequalities, equalities), and preferences (real-valued functions to maximize, minimize) either as real numbers or single real-valued distribution functions. Any large scale model of this type requires significant data gathering effort of precise values and/or judicious guessing. If the model has projections of future values, it is clear that real numbers and real-valued distributions are inadequate representations of parameters, even assuming that the model correctly captures the underlying system. It is also known from mathematical programming theory that only a few of the variables and constraints are necessary to describe an optimal solution (basic variables and active constraints) assuming a correct deterministic normative criterion, the objective function. Thus, only a few parameters need to be obtained precisely, those that are in the final basis. Of course the problem is that it is not known a-priori which variables will be basic and which constraints will be active.

Herbert Simon (see [25, 26]) states:

> Of course the decision that is optimal in the simplified model will seldom be optimal in the world. The decision maker has a choice between optimal decisions for an imaginary simplified world or decisions that are 'good enough,' that satisfice, for a world approximating the complex real one more closely. ... What a person cannot do he will not do, no matter how much he wants to do it. Normative economics has shown that exact solutions to the larger optimization problems of the world are simply not within reach or sight. In the face of this complexity the real-world business firm turns to procedures that find good enough answers to questions whose best answers are unknowable. Thus, normative microeconomics, by showing real-world optimization to be impossible, demonstrates that economic man is in fact a satisficer, a person who accepts 'good enough' alternatives, not because he prefers less to more but because he has no choice.

This observation is clearly reflected by Fandel, [31]. An E-mail discussion with Professor Rommelfanger [32] relates the following.

> In fact Herbert Simon develops a decision making approach, which he calls *the concept of bounded rationality*. He formulated the following two theses. *Thesis* 1: In general a human being does not strive for optimal decisions, but s/he tends to choose a course of action that meets minimum standards for satisfaction. The reason for this is that truly rational research can never be completed. *Thesis* 2: Courses of alternative actions and consequences are in general not known a-priori, but they must be found by means of a search procedure.

A model of an actual problem is always an abbreviated view of the underlying actual system. If a problem were able to be manipulated in its complex entirety in situ to obtain a solution without a symbolic representation, then there would be no need for modeling the problem mathematically in the first place. Inherently, a mathematical model is a symbolic representation of the problem not the problem itself. Thus, a model is in the process of knowing or discovery which we call *process epistemology*.

Mathematics is the science of precision. At the heart of analytical mathematics is order–order is more fundamental than measure (L. Collatz, lecture Oregon State University, Department of Mathematics lecture 1976). Measure (quantity and its generalizations) and extent (distance, area, integration) can be derived from order. Optimization requires an order, a normative criteria; it is the measure we impose, it is the objective function, and it is the objective function that tells us what is "best." A most ideal order is found in the set of real numbers (the real number line). In optimization we very often map, via the objective function, onto the real numbers. In flexible optimization, we map fuzzy sets (sets of membership functions) onto the set of real numbers using aggregation operators. Stochastic optimization often uses expected value to map distributions into the real number line. In optimization under generalized uncertainty (possibilistic optimization for example), we map pairs of distributions, for example possibility and necessity distributions, onto real numbers using an evaluation or scalarized [33] functional whose domain is that of the set of distributions and range is the set of real numbers or minimization of maximum regret (minimax optimization). For example, a generalized expectation is one such evaluation functional which takes as input functions and outputs a real number, an expectation which is a real-number.

Professor Roman Slowinski, however, once quoted, in articulating a differing point of view that the real numbers are an ideal order:

> Si l'ordre apparaît quelque part dans la qualité, pourquoi chercherions-nous à passer par l'intermédiaire du numbre?" Bachelard 1934. "If an order appears somewhere in quality, why should we like to interpret this order through numerical values?

Quality in this sense will not be our point of view and leave this discussion to others. Since our focus is quantitative entities, we will, for optimization, map into the real numbers.

A useful approach to fuzzy and optimization under generalized uncertainty should also adhere to the *Principle of Least Commitment* which states that, "Only commit when you must." One of the authors first heard the term, Principle of Least Commitment, used by Dr. J. Keller at a NAFIPS conference September 17–20, 1995, College Park, MD. In the context of optimization, the principle of least commitment is to use real-valued, deterministic, entities (defuzzification for fuzzy sets or expectations for distributions for example) only when a deterministic/crisp decision must be made, as close as possible to the last step in the modeling process. This is the approach that this text uses - to carry the full extent of uncertainty and/or gradualness from the beginning until one must deterministically choose as close as possible to the last step as illustrated by Example 15.

We summarize what has been articulated by quoting (and adding a few remarks) Dubois and Prade [34].

> * Fuzzy optimization, what we call here *flexible optimization*, offers a bridge between numerical, deterministic, approaches and the linguistic or qualitative ones. The thrust of these approaches are to provide the analyst with what is the uncertainty in the results of decision processes.
> * Fuzzy set theory and its mathematical environment of aggregation operators (*"and"*, *t-norms*), interval analysis, constrained interval analysis [35], fuzzy interval analysis, constrained fuzzy interval analysis, gradual numbers [36], and preference modeling, provide a general framework for posing decision problems in a more open way and provides a unification of existing techniques and theories.
> * Fuzzy set theory has the capacity of translating linguistic variables into quantitative terms in a flexible and useful way.
> * Possibility theory and generalized uncertainty theory explicitly accounts for lack of information, avoiding the use of unique, often uniform, probability distributions.
> * The set theoretic view of functions to represent numbers on which utilities (and scalarized functionals [33]) are expressed as fuzzy sets offer a wide range of aggregation operations.

In short, fuzzy set theory and generalized uncertainty theory offer optimization two approaches that are closest to the underlying flexible, satisficing, and information deficient processes that are very often the typical environment in which decision makers find themselves.

1.2.2   Fuzzy Set and Possibility Theory—Differences

It is important to clearly distinguish fuzzy sets from possibility not only since these concepts are, at times confused, but in optimization, fuzzy optimization and possibility/generalized uncertainty optimization have distinct methods of solution. Possibility theory grew out of fuzzy set theory and in fact, the title of the first paper on possibility theory was "Fuzzy sets as a basis for a theory of possibility," [37]. However, since its inception, it has been shown to be distinct from fuzzy set theory. So what distinguishes fuzzy sets and possibilities? First, as has been mentioned, fuzzy set theory is associated with generalized set theory. Possibility theory is associated with generalized probability theory we call generalized uncertainty which contains probability theory. Second, the semantics of fuzzy sets and possibilities are distinct. Fuzzy sets are associated with gradual set belonging and so its semantics is that of generalized sets. Possibilities are associated with information deficiencies and so its semantics are tied to generalized probability. Third, when looked at from first principles (axioms), the measures they induce (fuzzy measures, possibility measures) are distinct (see [38]). That is, the underlying measure theoretic foundations fuzzy set theory and possibility theory are distinct.

The confusion between fuzzy set theory and possibility theory arises most especially in optimization since both fuzzy optimization (flexible optimization) and possibility optimization (optimization under generalized uncertainty) use fuzzy intervals (numbers) which gives the impression that there is no distinction between fuzzy sets and possibilities. In particular, fuzzy intervals (numbers) may mathematically encode transitional set belonging and fuzzy intervals may also encode information deficiency, partially known values, and/or non-specificity. To insure that distinctions are maintained, one must understand the semantic of the underlying parameters and relationships, that is, the model. *It is crucial that theses distinctions are known since their corresponding optimization methods are different.*

The data from which the parameters and relationships arise will dictate its semantics. If the model encodes relaxed set belonging, it represents flexibility and its language is fuzzy set theory. If the model arises from data whose parameters are partially known, ambiguous, or incompletely specified, then it is a type of generalized uncertainty. Possibility theory is a mathematical language that may be used to mathematically represent this type of uncertainty. We will make precise, in Chap. 3, the construction method for the distributions from the entities' defining properties. There are several languages that capture generalized uncertainty, for example, random sets, Dempster-Shafer belief/plausibility, interval-valued probability, clouds. We discuss the association between possibility theory and these other languages subsequently since it will define the scope and limits of each of these mathematical languages as they apply to data used in optimization models. Because we show that the methods to solve flexible optimization and optimization under generalized uncertainty are distinct, it is crucial to know the nature of the underlying parameters and relationships. To this end, we introduce some basic concepts associated with these concepts.

1.2.2.1 Fuzzy Sets

A set is a fuzzy set if set belonging is gradual. As we mentioned, we take a very practical view of a real-valued fuzzy set $\tilde{A}$ as one that is uniquely defined by a continuous concave function $\mu_A(x)$, $x \in D \subseteq \mathbb{R} \to [0, 1]$, whose closure of its support is compact (a closed and bounded real interval) and whose range is a subset of $[0, 1]$. That is, for our exposition, a fuzzy set is comprised of two parts, (1) Its domain D, which is a subset of the real numbers $D \subseteq \mathbb{R}$, and a function we call the membership function, μ_A, that indicates the degree of belonging points of the domain possess. This is akin to a probability distribution of a random variable X which consists of a domain and a function on that domain, the probability distribution function that describes the probability on that domain. We assume the convention that for all membership functions,

$$\mu_A(x) = 0 \text{ if } x \in \emptyset.$$

An $\alpha - level\ set$ of a fuzzy set $\mu_A(x)$ for any $0 < \alpha \leq 1$ is defined to be,

$$A_\alpha = \{x | \mu_A(x) \geq \alpha\}. \tag{1.12}$$

Note that for continuous membership functions, the level sets are intervals [39]. There are generalizations to the assumptions made on the membership function such as upper semi-continuity rather than continuity, but we will not pursue these generalizations. Thus, for this monograph, we assume μ_A is continuous in which case its α-levels are intervals. This text does not consider discrete fuzzy sets.

What is important to note is that in the language of fuzzy sets, $\mu_A(x) = 1$ means that x belongs to the set A in the classical sense of belonging (does belong for sure) and $\mu_A(z) = 0$ means that z does not belong to set A in the classical sense of not belonging (does not belong for sure). All values between 0 and 1 denote the transition from belonging to not belonging and represent a relaxation of a Boolean set belonging.

1.2.2.2 Generalized Uncertainty

Generalized uncertainty as we use this term here denotes entities whose values are partially known. As we mentioned, we define generalized uncertainty, for this exposition, to be imprecisions whose representations are one of seven types: intervals, fuzzy intervals, possibility/necessity pairs, interval-valued probability distribution pairs, P-Boxes, clouds, and Kolmogorov, Smirnov statistics. There are in all at least ten uncertainty types of interest of which three uncertainty types (belief/possibility, probability intervals, imprecise probabilities, and random sets) are assumed to have sufficient hypotheses associated with the data to be transformed into interval-valued probability distributions. Thus, we focus on seven types which we consider as being fundamental. All of these types can be considered as interval-valued probabilities (see Fig. 2.1).

1.3   Fundamental Entities: Intervals, Fuzzy Intervals and Possibility Intervals

We next outline the relationship between fuzzy intervals and possibility distributions before looking at interval analysis as it relates to fuzzy interval analysis and possibility theory applied to optimization. We begin by looking at an example of what we mean by fuzzy (gradual set belonging) and possibility (lack of information).

Remark 16 For example, there is uncertainty in "the" minimum radiation dosage that will kill a cancer cell, as a unique real number, located at a particular pixel of a particular type of cancer for a particular person's computed tomography (CT) image. Suppose a radiation oncologist represents his/her uncertainty about the minimum radiation dosage as a distribution of preferred values (see Fig. 1.6, a trapezoid 58/59/61/62) where less than 58 definitely does not kill a cancer tumor cell while 62 is definitely more radiation than required to kill a tumor cell, with the range 59 to 61 being the most preferred and the interval most likely to contain the minimum radiation dosage to kill a cancerous cell. Here 58/59/61/62 (Fig. 1.6) encodes what is know about the minimal radiation level that will kill a cancer cell.

Remark 17 This fuzzy interval, the trapezoid 58/59/61/62, associated with "the" minimal radiation dosage that will kill the cancer cell, is a possibility distribution from which upper and lower bounding possibility measures can be constructed and used in optimization to enclose a set of distributions or singly as an uncertainty entity. In particular, when the trapezoidal fuzzy set is the encoding that represents the current state of our partial and incomplete knowledge, the trapezoid itself can model the fact that we do not know which distribution is the one which will be the one to be "the" minimal to kill the tumor cell and so we have an uncertainty. In the context of a CT image, the distribution taken from the uncertainty information encoded in the trapezoid 58/59/61/62,

$$\mu_{trap}(x) = \begin{cases} 0 \text{ for } x < 58, x > 62 \\ x - 58 \text{ for } 58 \leq x \leq 59 \\ 1 \text{ for } 59 < x < 61 \\ -x + 62 \text{ for } 61 \leq x \leq 62 \end{cases},$$

$$pos(x) = \begin{cases} 0 \text{ for } x < 58 \\ x - 58 \text{ for } 58 \leq x \leq 59 \\ 1 \text{ for } x > 59 \end{cases}$$

measures the **possibility** that the value x is likely to be a minimum dosage that will kill the cancer at the pixel in question. Of course, we are assuming that the trapezoid $\mu_{trap}(x)$ is the correct (uncertainty) information associated with the minimal tumor-cidal dose, that is, no dose below 58 has the ability to kill a tumor cell and doses above 62 are always more than minimal. Between these two values, the function can vary depending on the tumor (type, aggressiveness, ...), research results, oncologists's experience, and the particular patient characteristic. The distribution taken from the uncertainty information encoded in the trapezoid $\mu_{trap}(x)$

$$nec(x) = \begin{cases} 0 \text{ for } x < 61 \\ x - 61 \text{ for } 61 \leq x \leq 62 \\ 1 \text{ for } x > 62 \end{cases}$$

measures the **necessity** that the value x is likely to be a minimum dosage that will kill the cancer at the pixel in question. That is, nec(x) indicates the dosage that necessarily will kill a cancer tumor given the uncertainty in question and again assuming that the trapezoid $\mu_{trap}(x)$ indeed correctly encodes "minimum tumorcidal dose" So, for example, any radiation level 61 or above is guaranteed to kill the tumor cell.

Remark 18 Any fuzzy interval like $\mu_{trap}(x)$ generates three principle distributions —itself, its possibility, its necessity. From an optimization point of view, generalized uncertainties are characterized by: (1) A distribution that represents the state of partial knowledge, in our example, $\mu_{trap}(x)$; (2) A pair of distributions that enclose the possible distributions that can describe the actual situation, $pos(x)$ and $nec(x)$ in our example. This is the salient characteristic of what we call **generalized uncertainty**.

1.3.1 Intervals

An interval is a compact connected set of real numbers we denote by $[x]$ where $[x] = [\underline{x}, \overline{x}] = \{x | \underline{x} \leq x \leq \overline{x}\}$. There are approaches that allow one or both endpoints to be infinite defined on the extended real numbers, but we do not pursue this generalization. Hence,

$$-\infty < \underline{x} \leq \overline{x} < \infty$$

for this exposition. Intervals are key to the mathematical operations on fuzzy numbers, fuzzy intervals, and generalized uncertainty types in that the level sets of fuzzy intervals and the level sets of upper/lower possibility bounds of distributions are intervals. In particular, the usual way to compute with fuzzy intervals as generalized uncertainties is via intervals arising from level sets. More details are presented further in the text.

An interval $[a, b]$, from the point of view of flexibility, gradual set belonging, that is, a fuzzy set, may be considered as the following fuzzy set whose membership function is

$$\mu_{[a,b]}(x) = \begin{cases} 1 & x \in [a, b] \\ 0 & \text{otherwise} \end{cases} . \tag{1.13}$$

For example, if we have an interval $[1, 4]$, then as a fuzzy set whose membership function is $\mu_{[1,4]}(x)$ is depicted by Fig. 1.2.

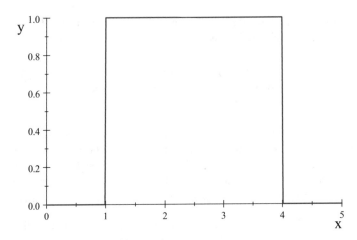

Fig. 1.2 Interval [1,4] fuzzy membership function

This set transitional belonging point of view, an interval, has two abrupt changes from not belonging ($x \notin [1, 4]$) and set belonging ($x \in [1, 4]$) represented by the vertical lines at $x = 1, x = 4$. Thus, an element either belongs to $[1, 4]$ or it does not. A more general fuzzy entity is a gradual set represented by a "sloped" left and/or right side of the fuzzy membership function which an interval is not (see Figs. 1.4, 1.5, Fortin, Dubois, Fargier [36] for further discussions). The arithmetic on fuzzy entities is fuzzy arithmetic (see Kaufmann and Gupta [40]) and the algebraic structure and mathematical space is that of fuzzy numbers (see [41]).

An interval can also encode non-specificity or information deficiency and is thus, in this context, possibilistic, a possibilistic interval. Given an interval $[a, b]$ from the point of view of representing probability-based possibility theory, if all we know is that the support of a probability is $[a, b]$, we can construct the following two pairs of (cumulative) distributions which are a pair of possibility distributions

$$pos(x) = \begin{cases} 0 \ x \in (-\infty, a) \\ 1 \ x \in [a, \infty) \end{cases}, \tag{1.14}$$

$$nec(x) = \begin{cases} 0 \ x \in (-\infty, b) \\ 1 \ x \in [b, \infty) \end{cases}. \tag{1.15}$$

These two functions, $pos(x)$ and $nec(x)$ are indeed possibility distributions with respect to its generating function $\mu_{[a,b]}$ (1.13) and have an important associated property that is true of every such type of generated $pos(x)$ and $nec(x)$. The important property is that if in fact, an interval, such as the one above, encodes an unknown probability distribution where the only information that is known is that the support of the probability distribution lies in $[1, 4]$, then **all** associated probability density functions $f(x)$ have cumulative distribution functions $F(x)$ on or between $pos(x)$

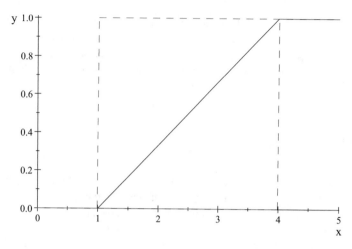

Fig. 1.3 Necessity ≤ uniform ≤ possibility

(1.14) and $nec(x)$. Equation (1.15). That is $F(x) \in [pos(x), nec(x)]$, or $pos(x) \leq F(x) \leq nec(x)$.

Suppose $[a, b] = [1, 4]$ for definiteness. The first function, $pos(x)$, (1.14), is an upper possibility distribution and the second function, $nec(x)$, (1.15), is a lower distribution. These are depicted in Fig. 1.3. Note that $pos(x)$ and $nec(x)$ are indeed cumulative probability distributions where $pos(x)$ corresponds to the case where the support is the singleton a and $nec(x)$ corresponds to the case where the support is the singleton b. *All* cumulative density functions whose corresponding probability density has support in $[1, 4]$ are functions that lie between the upper, dashed green (1.14) and lower, dashed red, (1.15) functions. That these are possibility distributions containing all cumulatives whose support lie in $[1, 4]$ will be derived in a more general setting subsequently. From the point of view of an interval representing probability-based possibility associated with incompleteness of information, ambiguous information, non-specificity, in the context of the interval $[1, 4]$, any distribution whose support lies in $[1, 4]$ is a valid instantiation of the given incomplete information represented and encoded by $\mu_{[a,b]}$. That is, the fuzzy interval $[a, b]$, which is the fuzzy interval $[1, 4]$ in this example, can encode an uncountably infinite family of cumulative distributions. One such cumulative is the one that corresponds to the uniform probability distribution (see Fig. 1.3) which is

$$\mu_{unif}(x) = \begin{cases} 0 \text{ for } x < 1 \\ \frac{1}{3}x - \frac{1}{3} \text{ for } 1 \leq x \leq 4 \\ 1 \text{ for } x > 4 \end{cases} .$$

We will see that these different representations depend on the context of the problem, the nature of the uncertainty. Intervals as an encoding of partial information, is one such representation of (generalized) uncertainty. The reader is already alerted to the fact that we are dealing with a different structure than that of the real (or complex) number system. In the above, we gave an example (Example 14) where $[r] = [s]$ is not equivalent to $[r] \leq [s]$ and $[r] \geq [s]$. When we deal with generalized uncertainty distributions, their algebraic properties are also different than those of real functions.

1.3.2   Fuzzy Intervals

What is usually called a fuzzy number is a particular case of a more general entity, a *fuzzy interval*. What is called a fuzzy number, say fuzzy number $\tilde{x}^*$, is a fuzzy set $\mu_{x^*}(x)$ for which $\mu_{x^*}(x^*) = 1$ uniquely. A fuzzy interval is the same except instead of a unique x^* for which $\mu_{x^*}(x^*) = 1$, for a fuzzy interval, where the membership is $\mu(x) = 1$, can be an interval $\underline{x} \leq x \leq \overline{x}$ with the possibility that $\underline{x} < \overline{x}$. A fuzzy interval can be translated into a pair of possibility distributions and thus may be a model for both the lack of specificity as well as gradualness. In particular, for optimization problems, fuzzy intervals can model both gradualness (flexibility) and uncertainty (lack of specificity). Thus, some fuzzy sets (fuzzy intervals, for example) *may* have or take on a double nature—that of capturing gradualness (flexibility) of belonging and capturing non-specificity, lack of information (uncertainty). When the same entity has multiple representations, as it does for fuzzy intervals, clear semantic distinctions are essential/required. Let us formalize what we mean by a fuzzy number and fuzzy interval.

Definition 19 A **fuzzy number** x^* is a fuzzy set over the domain of real numbers whose membership function is continuous with one and only one value, x^*, such that $\mu(x^*) = 1$ (x^* is the number that is "fuzzified") for which if $x \leq x^*$ then $\mu(x)$ is non-decreasing and if $x \geq x^*$ then $\mu(x)$ is non-increasing where its support is an interval containing x^*.

Definition 20 The set of domain elements (real numbers) for which the membership values are equal to one is called the **core**.

For example, a fuzzy interval $\tilde{2}$ has uniquely $\mu(2) = 1$, and may be depicted in Fig. 1.4. A fuzzy number, say a fuzzy $\tilde{2}$, very often arises in the context of obtaining a value for a parameter based on epistemic knowledge such as in, "the value is around 2." We denote a triangular fuzzy membership function by left endpoint/unique-core/right endpoint so that the fuzzy $\tilde{2}$ depicted in Fig. 1.4, is denoted 1/2/3.

Definition 21 A **fuzzy interval** $\tilde{M}$, depicted as a trapezoid in Fig. 1.5, is a fuzzy number except the core (which must also exist) does not have to be a singleton but may be a non-zero width interval.

We will denote a fuzzy trapezoid membership function by left endpoint/left-core endpoint/right-core endpoint/right endpoint so that the membership function

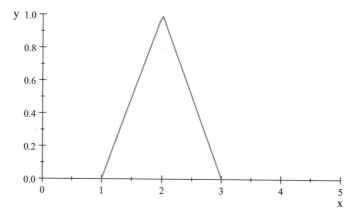

Fig. 1.4 Fuzzy interval 2

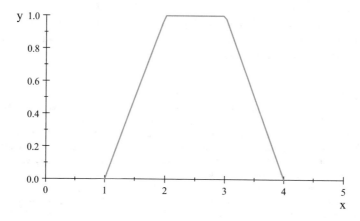

Fig. 1.5 Fuzzy trapezoidal interval

of Fig. 1.5 is denoted 1/2/3/4. All fuzzy numbers are fuzzy intervals so we will refer to any fuzzy set whose domain is the set of real numbers $\mathbb{R}$ with a non-empty core as a fuzzy interval. Moreover, intermingled with fuzzy sets and possibilistic or generalized uncertainty is the issue of the semantic designations *conjunctive* and *disjunctive* to entities, which arise in the *application* of fuzzy intervals. According to Dubois and Prade [42], fuzzy set theory and its extensions seem to come to grips with three basic concepts that noticeably differ from each other but may appear conjointly in various circumstances. The basic semantics of fuzzy and uncertainty include at least the following.

1. *Gradualness*: The idea that many categories in natural language are a matter of degree, including truth. The extension of a gradual predicate is a fuzzy set, a set where the transition between membership and non-membership is, in the words

of its inventor, "gradual rather than abrupt". This is Zadeh's original intuition [1].

2. *Epistemic uncertainty*: The idea of partial or incomplete information. In its most primitive form, it is often described by means of a set of possible values of some quantity of interest, one of which is the right one. This view leads to possibility theory [37].

3. *Ontic uncertainty:* The semantic associated with classical sets is called "*ontic*" according to [43]. Ontic means that, in the context of real number intervals, the interval is set of related elements, numbers, which form a *whole*, the interval.

4. *Bipolarity bipolar*: The idea that information can be described by distinguishing between positive and negative sides, possibly handled separately. For this text, we take bipolarity as being discrete states where this more general interpretation we call *polarity* or *polar.*

This text does not deal with ontic or bipolarity as an entity, only gradualness (fuzzy sets) and epistemic uncertainty (possibility, generalized uncertainty) Much more can be said about the topic of semantics. Only gradualness and epistemic are central to this presentation. However, to emphasize the distinction between these two semantics, since it is central to this presentation, we continue with what Dubois and Prade [42] go on to state:

As a mathematical notion, a fuzzy set F is unambiguously defined through a membership function μ_F from a set S to a (generally complete) lattice L. The mathematical object representing the fuzzy set is the membership function. Now, looking at the literature, it is clear that there are two ways of using a fuzzy set, which parallel the two ways sets are used. The fuzzy set is used as representing some precise gradual entity consisting of a collection of items. A typical example is a region in a grey image. Due to the presence of levels of grey, the boundary of a meaningful area in the image (e.g. forest zone or tumor in a CT image) is gradual. In many cases, it does not represent uncertainty about a precise boundary (just the fact that the density of trees or tumor cells is slowly decreasing in peripheral zones). Sometimes, the gradualness is linked to the use of a fuzzy predicate (e.g. the picture of a "populated area" or "tumor"). There is no uncertainty as soon as there is an agreement on the choice of the membership function (of "populated" or "tumor"). The same view of fuzzy sets is adopted for the gradual representation of clusters in data analysis, or linguistic partitions in fuzzy control or fuzzy random variables in the style of Puri and Ralescu [44], ranging on a set of (membership) functions. Similarly, the rating profile of a decision according to several criteria sharing a common value scale is of the same conjunctive nature. Such fuzzy sets are *conjunctive* and can be called *ontic* fuzzy sets. ... In contrast a fuzzy set may model incomplete information, namely be *disjunctive*. In that case, as pointed out by Zadeh [37], a membership function (with values on a totally ordered set) is then interpreted as a possibility distribution, denoted π over S. Now, S represents a universe of discourse, a frame of discernment, a set of possible worlds. This kind of fuzzy set is basically a subjective notion as it is attached to an agent whose epistemic state is described by distribution π.

Note: what is in parenthesis are the authors' clarifying insertions.

1.3.3 Possibility Intervals

Possibility theory was first proposed by Zadeh [37] and more extensively articulated in Dubois and Prade [27]. Necessity was first developed by Dubois and Prade [27]. In particular, if we know the possibility of a set A, the possibility of the complement of A, $Pos(A^C)$ is *not* defined as $1 - Pos(A)$ in contradistinction to probability. We need to define a concept that is the "dual" of possibility called necessity where

$$Nec(A) = 1 - Pos(A^C),$$

that is,

$$Pos(A^C) = 1 - Nec(A).$$

Dubois [34] states:

> Limited (minimal) specificity can be modeled in a natural way by possibility theory. The mathematical structure of possibility theory equips fuzzy sets with set functions, conditioning tools, notions of independence and dependence, decision-making capabilities [lattices]. Lack [deficiency] of information or lack of specificity means we do not have 'the negation of a proposition is improbable if and only if the proposition is probable.' In the setting of lack of specificity, 'the negation of a proposition is impossible if and only if the proposition is necessarily true.' Hence, in possibility theory pairs of possibility and necessity are used to capture the notions of plausibility [possibility] and certainty [necessity]. When pairs of functions are used we may be able to capture or model lack of information. *A membership function is a possibility only when the domain of a fuzzy set is decomposable into mutually exclusive elements.* A second difference [between probability and possibility besides possessing a dual necessity] lies in the underlying assumption regarding a probability distribution; namely, all values of positive probability are mutually exclusive. A fuzzy set is a conjunction of elements. For instance, in image processing, imprecise regions are often modeled by fuzzy sets. However, the pixels in the region are not mutually exclusive although they do not overlap. Namely the region contains several pixels, not a single unknown one. *When the assumption of mutual exclusion of elements of a fuzzy set is explicitly made, then, and only then, the membership function is interpreted as a possibility distribution; this is the case of fuzzy intervals describing the ill-located unique value of a parameter.* (here, the braces, [ ], indicate the authors' comments not in the original and the *italics* are the authors' emphases)

Moreover, possibility is normalized since the semantics of possibility is tied to an existential entity. That is, models that use possibility are of existential entities. *Thus, not all fuzzy set membership functions can generate possibility distributions.* So, again, it is crucial to distinguish gradualness from lack of information.

Pairs of functions can be considered a part of generalized uncertainty theory. When the data does not explicitly give us the associated membership pairs of functions that bound our lack of information, the data needs to be used to construct the bounding functions via methods we develop in Chap. 3 and each, when the upper and lower distributions are used, lead to a type of optimistic and pessimistic optimization, respectively, under the uncertainty model.

There are several ways to construct bounding distribution functions which will be more fully derived in the next chapter. The usual method is to construct bounding

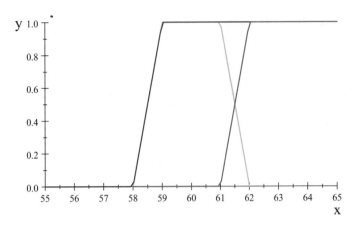

Fig. 1.6 Minimal tumorcidal dose

possibility distribution functions constructed from fuzzy intervals and illustrated by our example distributions generated by the interval [1, 4], depicted in Fig. 1.3 as well as $\mu_{trap}(x)$ priory and restated below. That is, the most prevalent approach is to define the entities of interest in optimization (the coefficients and/or the right-hand side values, for example) via fuzzy intervals in which case they will be able to model both gradualness and lack of specificity/information either as a single distribution or as pairs of distributions. In particular, suppose we have the trapezoidal fuzzy interval 58/59/61/62 depicted in Fig. 1.6. Then

$$\underline{pos}(x) = \begin{cases} 0 \text{ for } x < 58 \\ x - 58 \text{ for } 58 \le x \le 59 \\ 1 \text{ for } x > 59 \end{cases}$$

and

$$\overline{pos}(x) = \begin{cases} 0 \text{ for } x < 61 \\ x - 61 \text{ for } 61 \le x \le 62 \\ 1 \text{ for } x > 62 \end{cases}$$

define a pair of distributions that bound the uncertainty represented by the fuzzy interval 58/59/61/62.

More general fuzzy intervals can be constructed analogously. In particular, take the non-decreasing portion of the fuzzy interval such as 58/59/61/62 beginning at the left-most point of the support (58 in this case) of the domain of the interval until the first point at which the fuzzy interval is 1 (59 in this case, the leftmost core value) and set this non-decreasing function equal to $\underline{pos}(x)$, which in this case is $x - 58$, $x \in [58, 59]$. Then from that domain value until the right-most point of the domain the left or lower distribution $\underline{pos}(x) = 1$, which in this case is $x \in [59, +\infty)$. To the left from the leftmost point of the support of the fuzzy interval, set $\underline{pos}(x) = 0$, which in this case is $x \in (-\infty, 58]$. Next take the rightmost point of the core, which is 61 in our example and set $\overline{pos}(x) = 0$ from the smallest domain value to this rightmost core point, which means that $x \in (-\infty, 61]$. Then take the rightmost point of the support, which is 62 in our example, and take $1 - nonincreasing$ part of the fuzzy interval as the right or upper distribution, that is

$$\overline{pos}(x) = 1 - (-x + 62) = x - 61, x \in [61, 62],$$

for our 58/59/61/62 example. Next, from the rightmost point of the support, which is 62 in our example, to the rightmost point of the domain set the upper or right function to be 1 so that $\overline{pos}(x) = 1, x \in [62, +\infty)$. We formalize this construction subsequently.

Remark 22 We emphasize what Dubois and Prade [45] stated on page 77. "An expression such as 'X is F' where X is a variable and $\tilde{F}$ a fuzzy set (e.g.., 'age is young') may be used in two different types of situations, which both take advantage of the fuzziness of F. On one hand the expression 'X is F' can take place in which the value of X is precisely known and we can estimate the extent to which this value is compatible with the label F (whose meaning obviously depends on the context). In this case we are interested in the gradual or soft nature of the qualification stated by 'X is $\tilde{F}$'. ... However, the expression 'X is F' may in other situations mean 'all that is known about X is that X is F' (without knowing the value of X precisely in this case). This corresponds to the situation of incomplete information (pervaded with imprecision and uncertainty), a situation in which we can only order the possible values of X according to their level of plausibility or possibility. When a fuzzy set is used to represent what is known about the value of a single valued variable, the degree attached to a value expresses the level of possibility that this value is indeed the value of the variable. The fuzzy set $\tilde{F}$ is then interpreted as a possibility distribution [37], which expresses various shades of plausibility on the possible values of the ill-known variable X."

Concretely, in an optimization model, which seeks to kill cancerous tumors while sparing in the long run all other organs and tissues of a human body, if we have a constraint that is flexible—(a) "at a tumor pixel, deliver less than 61 units of radiation but under no circumstances deliver more than 62 units" and (b) "at a tumor pixel, deliver more than 59 units of radiation but never less than 58 units"—we have a fuzzy or flexible constraint. If we wish to deliver the minimal tumorcidal dosage represented by the fuzzy interval, the trapezoidal fuzzy number in Fig. 1.6, which was constructed from research and radiation oncology experience and is the state of knowledge we have about "minimal tumorcidal dose" as incomplete information about the minimal tumorcidal dose, then we have a possibility. While both the fuzzy or flexible set and possibilistic distribution are represented by the same function, these two representations lead to distinct optimization methods. The first (fuzzy/flexible) uses an aggregation operator and the second (possibilistic) uses a stochastic recourse-like approach.

A reason that one might want to use possibility rather than probability is precisely in situations for which real-values or complete probability density functions for data are not available. For example: (1) we do not know which of a given set of probabilities to use, or (2) all we know is that the probability is bounded by two functions, or (3) we do not have the probability distribution on singletons of the domain, but on subsets of the domain, some of which may have non-empty intersections (overlap). Whether an entity of interest,

- inherently lacks specificity (the minimal radiation that will kill a particular patient's prostate tumor cell located at (x, y, z) as a single real number), or
- lacks sufficient research to determine its precise value even assuming that a precise value such as "the minimum dosage" exists as a unique real number or as a precise probability density function, or
- does not need/require a deterministic functional representation, in the sense that one can get by with a more general form than its deterministic equivalent—perfect information, for the use to which it is put, for example, the light wave reflection measured by a satellite sensor to impute the depth of the ocean as low/medium/high might suffice for a given application such as the general morphology of the bottom of the ocean, or
- needs complexity reduction, for example, low, medium, high speed for the automatic gear shifting mechanism on a car,

lack of information/specificity is a part of many, dare we say most, problems. Moreover, when we have models that are built from epistemic knowledge (human ideas about the system rather than the system itself), many of these types of models are possibility either in their entirety or partially.

1.3.4　Interval, Fuzzy Interval, Representations, Arithmetic

A newer representation of intervals and fuzzy intervals is the constraint interval. These are defined next.

Definition 23　Constraint Interval (CI): is a representation [46] of the interval $[x] = \left[\underline{x}, \overline{x}\right]$ is a single-valued *function*

$$f(\lambda_x) = \underline{x} + \lambda_x \left(\overline{x} - \underline{x},\right) = \underline{x} + \lambda_x w_x, 0 \le \lambda_x \le 1, w_x = \overline{x} - \underline{x}.$$

Definition 24　Constraint Fuzzy Interval (CFI): is a CI representation of the alpha-levels of a fuzzy interval l

$$A(\alpha_A) = \left[\underline{x}(\alpha_A), \overline{x}(\alpha_A)\right], 0 \le \alpha_A \le 1,$$

of a fuzzy interval membership function. It is a bi-variable function

$$f(\alpha_A, \lambda_{x,A}) = \underline{x}(\alpha_A) + \lambda_{x,A}(\alpha_A)\left(\overline{x}(\alpha_A) - \underline{x}(\alpha_A),\right) = \underline{x}(\alpha_A) + \lambda_{x,A}(\alpha_A)w_{xA},$$
$$0 \leq \lambda_{x,A}(\alpha_A), \alpha_A \leq 1, w_{x,A} = \overline{x}(\alpha_A) - \underline{x}(\alpha_A).$$

The arithmetic associated with these representations are just the real function arithmetic

$$f(\lambda_x) * f(\lambda_y), * \in \{+, -, \times, \div\},$$
$$f(\alpha_A, \lambda_{x,A}) * f(\alpha_B, \lambda_{y,B}), * \in \{+, -, \times, \div\}.$$

There are advantages to this representation since inverses exist which is not true of the traditional interval and fuzzy interval arithmetic (see [41]). We do not develop these ideas further. The interested reader is directed to the various publications of the bibliography, in particular [47].

1.4 A Taxonomy of Generalized Uncertainty in Optimization

This exposition considers optimization problems to be (1.6)–(1.8) in the presence of data $\{a, b, c, d, e\}$ that are either all or a mixture of real, interval, interval-valued probability, possibility pairs, that is, generalized uncertainty, with at least one parameter being one of these types or relationships $\{\leq, =, \in\}$ that are flexible or soft. Next, a taxonomy is developed, since the taxonomy is related to solution methods in the chapters that follow. This will be discussed in greater detail the items that are presented next. Optimization in the presence of generalized uncertainty coefficients, and flexibility may be considered to be the following.

1. **Flexible Optimization**

    (a) **Soft constraints relationships**: The mathematical relationships $\leq$, $=$, and/or $\in$ take on a flexible meaning. For example, we may have a soft "less than or equal to" constraint where its meaning is: "I wish to be less than a certain value but I can tolerate going over the value a 'bit' beyond what is wished but certainly not more than this 'bit'." The extent of the flexibility will need to be specified. Another type of flexibility is a specification of being greater than a given probability, possibility, or necessity of a constraint violation. These latter flexibilities are chance-like constraints called modalities in the case of possibility and necessity (see [48]).

    (b) **The objective function expresses an acceptable set of desired targets**: A decision might be to attain an ideal amount of profit that is only attained under a rare confluence of situations, for example, it happened only once before. A flexible object function would be to come as close to the ideal as possible or exceed this ideal. These models include the original Bellman and Zadeh model [49] and Zimmermann's model [50]. The objective is stated as obtaining or exceeding, in the case of maximization, or coming under, in the case of a minimum, one or more targets that are fixed as a number, somehow computed, or subjectively chosen. These objectives are added to the set of constraint inequalities and treated as a soft constraint.

(c) **The right-hand-side value of a constraint is a fuzzy interval which is semantically a target:** The goal for a constraint might be, "Do not deliver less than 58 units of radiation to the tumor, preferably between 59 units and 61 units, but never exceed 62 units." When the fuzzy interval is semantically a target, it is a flexible constraint resulting in a flexible optimization problem. Such a right-hand side value is essentially a flexible relationship that is described by trapezoidal fuzzy interval in this example. These are translated into two constraints

$$A_{tumor,\circ}x \geq 58 + \alpha$$
$$A_{tumor,\circ}x \leq 62 - \alpha$$

where we maximize $0 \leq \alpha \leq 1$, where x is the radiation level coming from the radiation unit (linear accelerator), A is the radiation attenuation model and the subscript $tumor, \circ$ is the coordinate of the tumor voxel. This means that $A_{tumor,\circ}x$ is model of the dose delivered to a tumor voxel (see [51])

2. **Optimization Under Generalized Uncertainty**—With a single type of generalized uncertainty in the constraint set definition we have the following.

   (a) **Distributions on Constraint Set Parameters:** We consider five different generalized uncertainty types associated with the constraint set.
      (i) The first is when a fuzzy interval is used to capture the uncertainty of a set of parameters. The trapezoid depicted in Fig. 1.6 is a typical fuzzy interval constraint. Suppose data regarding the exact minimal radiation dose that will kill a cancerous cell comes from information that is partial or not specific. One expert says 58 is the minimum tumorcidal dose, another 62, another 59, another 60, and another 61. All experts, experimental results, experiential data, and evidence indicate that less than 58 units will certainly not kill the cancerous cell and the minimum radiation is certainly not more than 62 units. Most experts and evidence indicate that the best range is between 59 and 61. One model for this incomplete, non-specific, or deficient information is the trapezoidal fuzzy interval 58/59/61/62 itself. However, this is a possibilistic distribution and semantically distinct from the right-hand side goal stated in classification 1(c) above. While the trapezoid is the same and the location within the constraint sets (right-hand side) is the same, how we handle fuzzy and generalized uncertainty is distinct, which will be seen.
      (ii) The second is to use the fuzzy interval to generate another "upper cumulative" which is the lower possibility distribution, $\underline{pos}(x)$, rep-

resenting the most optimistic view of the values of the fuzzy interval (see the black solid line in Fig. 1.6, which is a cumulative possibility and the most optimistic scenario given the fuzzy interval parameter depicted).

(iii) The third is to generate the "lower cumulative" (upper) possibility distribution from the fuzzy interval, $\overline{pos}(x)$, representing the most pessimistic/conservative view of the values of the given fuzzy interval (see the red solid line in Fig. 1.6, which is a cumulative necessity and it is the most pessimistic scenario given the trapezoidal fuzzy interval parameter depicted).

(iv) The fourth is any distribution that is between the possibility and necessity. For example, one might choose for a trapezoidal parameter depicted by Fig. 1.6, the "midpoint" distribution between the upper and lower possibilities, that is, the uniform one, which is the solid purple one in Fig. 1.6.

(v) The fifth is really (ii) and (iii) combined simultaneously in a way to obtain something like a minimax regret (see [52, 53]) or penalized optimization (see [28, 29, 54–56]). That is, consider the pessimistic, necessity or lower distribution, and the optimistic, possibility or upper distribution, together in a way that incorporates the lack of information represented by the fuzzy trapezoidal interval in Fig. 1.6, for example. This type of distribution optimization is used when one wants to analyze and incorporate the risk of taking a particular course of action in the presence of the lack of information represented by the fuzzy interval. These methods give upper (optimistic), lower (pessimistic), and minimax regret. The difference between the optimistic and pessimistic solution can be used as a measure of the maximum risk in taking a particular course of action.

(b) **Distributions on Objective Function Coefficients**: Suppose the objective function parameters $\vec{c}$ are semantically linked to information deficiency, for example, the cost of travel time is uncertain in an optimization problem which minimizes fuel costs. This case is distinct from the case where the objective function is considered as a goal and translated into a soft constraint. For generalized uncertainty cost coefficients, we have an uncountably infinite family of possible objective functions. However, this family is bounded above and below, that is, the family of objective functions are contained in an envelope so we have an infinite multi-objective set of functions bounded above and below. Given the bounds, an optimistic, pessimistic, and minimax regret may be obtained or an interval expectation, to be derived, can be computed.

3. **Mixed Optimization**

(a) **Flexibility and Uncertainty: Single Type per Constraint Relation—** When a given constraint entirely is flexible (soft) and another constraint is entirely described by a generalized uncertainty type, one aggregates (to be defined later) the soft constraints and uses either chance constraints or penalized distribution. The objective function, which may now be a family of functions, is the input into a scalarized functional (see [33, 57] for example) such as a generalized expected average, defined below. To the best of our knowledge, mixed problems were first treated by Inuiguchi, Ichihashi and Tanaka [2].

(b) **Uncertainties of More Than One Type in One Constraint Relationship**—When any of the coefficients a, b, c, d, e of our deterministic model (1.6)–(1.8) appear as a mixture of generalized uncertainty types within the same constraint statement we have a second type of mixed optimization. For example, in one constraint relation, one coefficient might be a random set and another coefficient in that same relation might be an interval-valued probability. In this case, we may transform them into interval-valued probability distributions (see [56] and Fig. 2.1) since interval-valued probability (see Fig. 2.1) is the general type of generalized uncertainty that includes the other types as special cases. That is, random sets can (under certain conditions) be transformed into an interval-valued probability. In particular, when we do this, we must map all uncertainty types into the most general uncertainty type that include all uncertainty types present in a given constraint relations as subsets. In addition, we must develop a new extension principle to handle pairs of distributions (and all distributions between the pairs) since the integral of the generalized uncertainty distribution function needs to be evaluated in computing generalized expectations. These new extension principles will be discussed in Chap. 2 and methods to obtain functions of these generalized uncertainty types will be developed. A new type of a generalized uncertainty integration is needed if these multiple generalized uncertainty types occur in coefficients of the objective function and/or we need to use an evaluation functional something like the expected average/value, which maps distributions to the real numbers $\mathbb{R}$. This has been done (see [56]). More general mixed forms can also be handled by similar methods, see [53].

1.5 Summary

This chapter delineated the compelling strengths of flexible and generalized uncertainty optimization. It looked at the differences between flexible and generalized uncertainty optimization and indicated how the semantics are important in distinguishing which of the various approaches must be used in seeking a solution to associated models. The taxonomy of types of optimization that are linked flexible and generalized uncertainty optimization were indicated. Clearly, flexible and generalized uncertainty optimization problems are relevant in applications. And if we are to believe H. Simon, normative mathematical models (optimization in our case) are a process of discovery closely tied to process epistemic knowledge about uncertainty in the data or model itself. Much epistemic knowledge can be mathematized as fuzzy sets and fuzzy sets into deterministic equivalent models to be solved. Moreover, when we have lack of information in the data, the resulting uncertainty can be translated into possibility and necessity bounds where Chap. 3 presents approaches to construct upper and lower bounding functions for these uncertainties. Once we have bounding functions, it is, in principle, straight forward to translate the problem into an optimistic and pessimistic deterministic optimization problem. In fact, we have three:

1. The fuzzy interval itself or the fuzzy interval generated by the bounding functions, which will generate either a soft constraint optimization or a surprise optimization;
2. Any distribution between the bounding distributions including the bounding functions themselves;
3. Both distributions that are used in a min/max optimization.

1.6 Exercises

Exercise 25 Given the triangular fuzzy number 20/30/40, provide the graph of its corresponding upper possibility and lower necessity functions that enclose the lack of information represented by this fuzzy number.

Exercise 26 Read [58, 59] and then classify these approaches according to our taxonomy.

Exercise 27 Give another example where $[r] = [s]$ does not imply that $[r] \leq [s]$ and $[r] \geq [s]$. Explain mathematically why is this occurs for intervals and not for real-numbers?

References

1. L.A. Zadeh, Fuzzy sets. Inf. Control **8**, 338–353 (1965)
2. M. Inuiguchi, H. Ichihashi, H. Tanaka, Fuzzy programming: a survey of recent developments, in *Stochastic versus Fuzzy Approaches to Multiobjective Mathematical Programming under Uncertainty*, ed. by R. Slowinski, J. Teghem (Kluwer Academic Publishers, Dordrecht, 1990), pp. 45–68
3. D. Dubois, H. Prade, Formal representations of uncertainty, in *Decision-Making Process*, ed. by D. Bouyssou, D. Dubois, H. Prade (ISTE, London& Wiley, Hoboken, 2009)
4. R.E. Moore, *Methods and Applications of Interval Analysis* (SIAM, Philadelphia, 1979)
5. R.E. Moore, R.B. Kearfott, M.J. Cloud, *Introduction to Interval Analysis* (Society for Industrial and Applied Mathematics, Philadelphia, 2009)
6. D. Dubois, H. Prade, Random sets and fuzzy interval analysis. Fuzzy Sets Syst. **42**(2), 1987, 87–101 (1991)

7. D. Dubois, E. Kerre, R. Mesiar, H. Prade, Fuzzy interval analysis, in *Fundamentals of Fuzzy Sets*, ed. by D. Dubois, H. Prade (Kluwer Academic Press, 2000), pp. 483–581
8. G.J. Klir, B. Yuan, *Fuzzy Sets and Fuzzy Logic: Theory and Applications* (Prentice Hall, New Jersey, 1995)
9. K. Weichselberger, The theory of interval-probability as a unifying concept for uncertainty. Int. J. Approx. Reason. **24**, 149–170 (2000)
10. S. Ferson, V. Kreinovich, R. Ginzburg, K. Sentz, D.S. Myers, *Constructing Probability Boxes and Dempster-Shafer Structures*. Sandia National Laboratories, Technical Report SAND2002-4015, Albuquerque, New Mexico (2003)
11. S. Destercke, D. Dubois, E. Chojnacki, Unifying practical uncertainty representations: I. generalized p-boxes. Int. J. Approx. Reason. **49**, 649–663 (2008)
12. A. Neumaier, Structure of clouds (2005) (downloadable http://www.mat.univie.ac.at/~neum/papers.html)
13. S. Destercke, D. Dubois, E. Chojnacki, Unifying practical uncertainty representations: II. clouds. Int. J. Approx. Reason. **49**, 664–677 (2008)
14. A.N. Kolmogorov, Confidence limits for an unknown distribution function. Ann. Math. Stat. **12**, 461–463 (1941)
15. A.P. Dempster, Upper and lower probability induced by a multivalued mapping. Ann. Math. Stat. **38**, 325–339 (1967)
16. G. Shafer, *A Mathematical Theory of Evidence* (Princeton University Press, Princeton, 1976)
17. L.M.D. Campos, J.F. Huete, S. Moral, Probability intervals: a tool for uncertain reasoning. Int. J. Uncertain. Fuzziness Knowl.-Based Syst. **2**(2), 167–196 (1994)
18. H. Nguyen, *An Introduction to Random Sets* (Chapman & Hall/CRC, Boca Raton, 2006)
19. P. Walley, *Statistical Reasoning with Imprecise Probabilities* (Chapman & Hall, London, 1991)
20. J. Mendel, R.I. John, Type 2 fuzzy sets made simple. IEEE Trans. Fuzzy Syst. **10**(2), 117–127 (2002)
21. J. Ramik, M. Vlach, *Generalized Concavity in Fuzzy Optimization and Decision Analysis* (Kluwer Academic Publishers, Boston, 2002)
22. D.M. Gay, Solving linear interval equations. SIAM J. Numer. Anal. **19**(4), 858–870 (1982)
23. C.V. Negoita, The current interest in fuzzy optimization. Fuzzy Sets Syst. **6**, 261–269 (1982)
24. S.A. Orlovsky, On formalization of a general fuzzy mathematical problem. Fuzzy Sets Syst. **3**, 311–32 (1980)
25. H.A. Simon, *A New Science of Management Decision* (Harper, New York, 1960)
26. H.A. Simon, *Models of Bounded Rationality* (MIT Press, Cambridge, Mass, 1997)
27. D. Dubois, H. Prade, *Possibility Theory* (Plenum Press, New York, 1988)
28. K.D. Jamison, Modeling uncertainty using probability based possibility theory with applications to optimization. Ph.D. Thesis, UCD Department of Mathematics (1998)
29. K.D. Jamison, W.A. Lodwick, Fuzzy linear programming using penalty method. Fuzzy Sets Syst. **119**, 97–110 (2001)
30. H.J. Rommelfanger, The advantages of fuzzy optimization models in practical use. Fuzzy Optim. Decis. Mak. **3**, 295–309 (2004)
31. G. Fandel, *PPS-Systeme: Grundlagen, Methoden, Software, Mark-analyse* (Springer, Heidelberg, 1994)
32. H.J. Rommelfanger, Personal communication (2009)
33. M. Ehrgott, *Multicriteria Optimization* (Springer Science & Business, 2006)
34. D. Dubois, The role of fuzzy sets in decision sciences: old techniques and new directions. Fuzzy Sets Syst. **184**(1), 3–28 (2010)
35. W.A. Lodwick, Fundamentals of interval analysis and linkages to fuzzy set theory, in *Handbook of Granular Computing*, ed. by W. Pedrycz, A. Skowron, V. Kreinovich (Wiley, West Sussex, England, 2008), pp. 55–79
36. J. Fortin, D. Dubois, H. Fargier, Gradual numbers and their application to fuzzy interval analysis. IEEE Trans. Fuzzy Syst. **16**, 388–402 (2008)
37. L.A. Zadeh, Fuzzy sets as a basis for a theory of possibility. Fuzzy Sets Syst. **1**, 3–28 (1978)

38. M.L. Puri, D. Ralescu, Fuzzy measures are not possibility measures. Fuzzy Sets Syst. **7**, 311–313 (1982)
39. H. Nguyen, On conditional possibility distributions. Fuzzy Sets Syst. **1**(4), 299–309 (1978)
40. A. Kaufmann, M.M. Gupta, *Introduction to Fuzzy Arithmetic - Theory and Applications* (Van Nostrand Reinhold, 1985)
41. W.A. Lodwick, O. Jenkins, Constrained intervals and interval spaces. Soft Comput. **17**(8), 1393–1402 (2013)
42. D. Dubois, H. Prade, Gradualness, uncertainty and bipolarity: making sense of fuzzy sets. Fuzzy Sets Syst. **192**, 3–24 (2012)
43. I. Couso, D. Dubois, Statistical reasoning with set-valued information: Ontic vs. epistemic views. Int. J. Approx. Reason. (2013). https://doi.org/10.1016/j.ijar.2013.07.002
44. M.L. Puri, D. Ralescu, Fuzzy random variables. J. Math. Anal. Appl. **114**, 409–422 (1986)
45. D. Dubois, H. Prade, (eds.) *Fundamentals of Fuzzy Sets* (Kluwer Academic Press, 2000)
46. W.A. Lodwick, Constrained Interval Arithmetic, CCM Report **138** (1999)
47. W.A. Lodwick, Interval and fuzzy analysis: an unified approach, in *Advances in Imagining and Electronic Physics*, ed. by P.W. Hawkes, vol. 148 (Academic, 2007), pp. 75–192
48. M. Inuiguchi, H. Ichihashi, Y. Kume, Relationships between modality constrained programming problems and various fuzzy mathematical programming problems. Fuzzy Sets Syst. **49**, 243–259 (1992)
49. R.E. Bellman, L.A. Zadeh, Decision-making in a fuzzy environment. Manag. Sci. Ser. B **17**, 141–164 (1970)
50. H. Zimmermann, Description and optimization of fuzzy systems. Int. J. Gen. Syst.**2**, 209–215 (1976)
51. W.A. Lodwick, K. Bachman, Solving large scale fuzzy possibilistic optimization problems. Fuzzy Optim. Decis. Mak. **4**(4), 257–278 (2005)
52. P. Thipwiwatpotjana, W. Lodwick, The use of interval-valued probability measures in fuzzy linear programming: A constraint set approach, in *Proceedings of IFSA-EUSFLAT 2009*, Lisbon, Portugal (2009). Accessed 20–24 July 2009
53. P. Thipwiwatpotjana, *Linear programming problems for generalized uncertainty*, Ph.D. Thesis. University of Colorado, Department of Mathematical and Statistical Sciences (2010)
54. W.A. Lodwick, K.D. Jamison, Interval methods and fuzzy optimization. Int. J. Uncertain. Fuzziness Knowl.-Based Reason. **5**(3), 239–250 (1997)
55. W.A. Lodwick, K.D. Jamison, Interval-valued probability in the analysis of problems that contain a mixture of fuzzy, possibilistic and interval uncertainty, in *2006 Conference of the North American Fuzzy Information Processing Society, June 3-6 2006, Montréal, Canada*, ed. by K. Demirli, A. Akgunduz. paper 327137 (2006)
56. W.A. Lodwick, K.D. Jamison, Interval-valued probability in the analysis of problems containing a mixture of possibility, probabilistic, and interval uncertainty..Fuzzy Sets Syst. **159**(1), 2845–2858. Accessed 1 Nov 2008
57. J. Jahn (ed.) *Vector Optimization* (Springer, Berlin, 2009)
58. H.R. Maleki, M. Tata, M. Mashinchi, Linear programming with fuzzy variables. Fuzzy Sets Syst. **109**, 21–33 (2000)
59. M.R. Safi, H.R. Maleki, E. Zaeimazad, A geometric approach for solving fuzzy linear programming problems. Fuzzy Optim. Decis. Mak. **6**, 315–336 (2007)

Chapter 2
Generalized Uncertainty Theory: A Language for Information Deficiency

2.1 Introduction

The generalized uncertainty to which we restrict ourselves, as we have mentioned, are those uncertainties that are more general than those that are deterministic or single probabilistic distributions. After the theoretical aspects of generalized uncertainty are reviewed, this chapter shows how to construct distributions, which will be used as inputs to optimization under generalized uncertainty models. The generalized uncertainties to which this text limits itself are those that either are or can be translated into pairs of bounding distributions or interval-valued probabilities, since these two types seem to us to be most useful in optimization. They encompass all the uncertainty types of interest, and are the most directly applicable. Thus, our primary focus in this chapter is on generalized uncertainty types that lead to upper and lower distributions which are used in our generalized uncertainty optimization approach.

It is clear that many researchers consider fuzzy sets as uncertainty even though a fuzzy set is *uniquely* defined by a single membership function, that is, we have complete information. Generalized uncertainty, as the term is used here, arises from *partial* information so, in our usage of the term "uncertainty", fuzzy sets and uncertainty are distinct. Our definitions of these terms make this clear albeit it is not necessarily how these terms are used in the literature.

One of the most fundamental entities in mathematics is, arguably, that of set. When an application represents its entities as set, the application imposes attributes onto the set beyond the notion of a collection of objects. This arises from the nature of the application and/or the input data used in the application. For example, we may have a set of numbers, say, $\{2, ..., 12\}$, that represent the states of a pair of dice on which we can impose a probability. For the set $\{2, ..., 12\}$, the probability attached to the elements of the set is an induced attribute associated with the use of this set in modeling the states of dice. In this case we attach a function, the probability density function, to the elements of the set. In the case of a fuzzy set, we also have induced attributes, its associated grade of belonging to the set, which is also a function, a membership function, that is attached to the elements of a set.

© Springer Nature Switzerland AG 2021
W. A. Lodwick and L. L. Salles-Neto, *Flexible and Generalized
Uncertainty Optimization*, Studies in Computational Intelligence 696,
https://doi.org/10.1007/978-3-030-61180-4_2

Other attributes may also be attached such as an "or" or an "and". For example, when it is stated that an individual in a mathematics class is "tall", individual students in a particular mathematics class may be considered as tall if a person looking at them evaluates them as falling in the set {6'0" or ... or 7'0"}. Here "or" (disjunction) is attached to the set of height measurements of students in a particular class. On the other hand, if one looks at a set of pixels of a digital image of a tropical forest and one is singling out the set of pixels that form a particular "tree", it is clear that the tree in question is made up collectively of a group of pixels. Here "and" (conjunction) is attached to the set of pixels (pixel$_j$ *and* pixel$_k$ *and* ...) associated to the set "tree". "Tree" is well-defined since the person knows which tree. However, the pixels that belong to the person's "tree" may not. That is, whether or not a pixel belongs to "tree", the one in question, is a matter of transition as the pixels blend near the edges. Regardless, *and* is imposed on the set identified as "tree".

The point we wish to make is that attention and care to the attached attributes is required since these attributes that are implicitly or explicitly attached to the primary entity, the set, will make a difference on how the entity is handled mathematically. The modeling or use to which one puts sets will determine the associated attributes and properties of the set that are induced, which, in the cases we consider, will make a difference in how we are able to compute with the sets. Dubois and Prade [3] classify conjunctive as *ontic* and disjunctive as *epistemic*. This was introduced in Chap. 1. Loosely speaking, from our point of view, epistemic will be associated with information deficiency leading to one of our generalized uncertainty types. We will make these distinctions as we proceed.

Generalized uncertainty types begin with set-valued functions. Set-valued functions are functions whose domains are sets, generally not singleton sets. Regardless, the domain of set-valued functions are set, which require an underlying set structure that will allow us to construct these upper/lower bounding distribution functions. Set-valued functions with some hypotheses such as non-negativity and perhaps (sub)additivity are also called measures. We begin with probability measures which are well-known concepts from measure theory. The domain, a subset of sets, need a structure and for our purposes, the set structure we use is a sigma algebra, which we discuss next. Note that in this text by a **partition** of a set X we will mean an at most countable collection of pair-wise disjoint subsets whose union is X.

Definition 28 Given a (universal) set $X \neq \emptyset$, a **sigma algebra** defined on X, denoted σ_X, is a family of subsets of X such that:

1) $\emptyset \in \sigma_X$;
2) $X \in \sigma_X$;
3) $A \in \sigma_X \Rightarrow A^C \in \sigma_X$, A^C the complement of A;
4) $A_i \in \sigma_X$, for any countable set (could be finite) $\Rightarrow \{\cup_i Ai\} \in \sigma_X$.

The pair (X, σ_X) is called a **measurable space**. Note that the power set of X, $PS(X)$, is a sigma algebra. Let (X, σ_X) be a measurable space. By a measure μ on this space we mean a set-valued function

$$\mu : \sigma_X \rightarrow \mathbb{R}^+$$

such that $\mu(\emptyset) = 0$ and for any partition of X, $A_i \in \sigma_X$, $\mu(\cup_i A_i) = \sum_i \mu(A_i)$. The triple (X, σ_X, μ) is called a **measure space**. If the mapping is $\mu : \sigma_X \to [0, 1]$ has the property that $\mu(X) = 1$, then the measure is called a **probability measure** with μ now denoted $\Pr_X$ (where we drop the subscript when the context is clear) and the measure space is called a **probability measure space** denoted $(X, \sigma_X, \Pr_X)$. Let (X, σ_X) and (Y, σ_Y) be two measurable spaces. A function

$$f : X \to Y$$

is said to be (σ_X, σ_Y) **measurable** if

$$f^{-1}(A) = \{x \in X \mid f(x) \in A\} \in \sigma_X \text{ for each } A \in \sigma_Y.$$

In many situations there is insufficient information to accurately construct an underlying probability measure. We use a relaxed definition [4] to address these situations. Let $Int_{[0,1]} = \{[a, b] \mid 0 \le a \le b \le 1\}$.

Definition 29 Given measurable space (X, σ_X), then $i_m : \sigma_X \to Int_{[0,1]}$ is called an **interval-valued probability measure (IVPM)** on (X, σ_X) if it satisfies the following:

(1) $i_m(\phi) = [0, 0]$
(2) $i_m(\infty) = [1, 1]$
(3) $\forall A \in \sigma_X, i_m(A) = [A^l, A^u] \subseteq [0, 1]$
(4) for every partition of X, $\{A_{k \in K}\}$, $\{B_{j \in J}\} \subseteq \sigma_X$ such that $A = \cup_{k \in K} A_k$ and $A^c = \cup_{j \in J} B_j$ then

$$i_m(A) \subseteq \begin{bmatrix} \max\left\{1 - \Sigma_{j \in J} B_j^u, \Sigma_{k \in K} A_k^l\right\}, \\ \min\left\{1 - \Sigma_{j \in J} B_j^l, \Sigma_{k \in K} A_k^u\right\} \end{bmatrix}$$

We call (X, σ_X, i_m) an **interval-valued probability measure space**.

One of the central themes of this book is that almost all of the representations used to model data insufficiency in this text can be shown to be equivalent to an IVPM. This allows us to put models with different representations into one representation and solving problems using IVPMs applies to multiple settings. Typically, for functions of sets of real numbers, the smallest sigma algebra generated by the half-open sets of real numbers, $[a, b), a < b$, (called the *Borel* sigma algebra, $\mathcal{B}$) is used. As mentioned, the power set of any subset of $\mathbb{R}$ is a sigma algebra containing all other sigma algebras so that by intersecting all sigma algebras that contain the set of subsets generated by half open sets, we will have the Borel sets which will be the structure we need.

Remark 30 To see the motivation behind the above definition of an IVPM suppose we have an interval valued measure space $(\mathcal{R}, \mathcal{B}, i_m)$ which provides a model for an unknown random variable X such that $\forall A \in \mathcal{B}$ we know that $\Pr(X \in A) \in i_m(A)$. Let A, B, C, D be mutually disjoint with union all of $\mathcal{R}$ (i.e. they are a partition of $\mathcal{R}$). Consider $\Pr(X \in A \cup B)$. Since these sets are disjoint, the maximum this probability can be is $A^u + B^u$, the maximum probabilities of A and B separately. Similarly, the minimum is $A^l + B^l$. Combined we have

$$\Pr(X \in A \cup B) \in \left[A^l + B^l, A^u + B^u\right]$$

Similarly we have

$$\Pr(X \in C \cup D) \in \left[C^l + D^l, C^u + D^u\right]$$

But since $C \cup D = (A \cup B)^c$ and $\Pr(X \in A \cup B) = 1 - \Pr\left(X \in (A \cup B)^c\right)$ we know that

$$\Pr(X \in A \cup B) \in \left[1 - \left(C^u + D^u\right), 1 - \left(C^l + D^l\right)\right]$$

Combining these two bounds on $\Pr(X \in A \cup B)$ gives

$$\Pr(X \in A \cup B) \in \left[\begin{array}{c} \max\left\{1 - \left(C^u + D^u\right), A^l + B^l\right\}, \\ \min\left\{1 - \left(C^l + D^l\right), A^u + B^u\right\} \end{array}\right]$$

which is consistent with our definition for an IVPM. Thus for any $A \in \mathcal{B}$ the interval $i_m(A)$ is as tight as possible.

Example 31 Let $X \subset \mathbb{R}$ with partition A, B, C, D (mutually disjoint whose union is X) and $\quad i_m(A) = [0.2, 0.5], i_m(B) = [0.1, 0.15].i_m(C) = [0.15, 0.2], i_m(D) = [0.3, 0.4]$.

$$i_m(A) \subseteq [\max\left\{\left(1 - (B^u + C^u + D^u)\right), A_l\right\}, \min\left\{\left(1 - (B^l + C^l + D^l)\right), A^u\right\}$$
$$= [\max\{(1 - (0.15 + 0.2 + 0.4)), 0.2\}, \min\{(1 - (0.1 + 0.15 + 0.3)), 0.5\}]$$
$$= [\max\{0.25, 0.2\}, \min\{0.45, 0.5\}] = [0.25, 0.45].$$

That is, the bounds on $i_m(A)$ were tightened. The uncertainty in A is related to the uncertainty in A^c which means that the uncertainty on collected interval data can be used to reduce the uncertainties in the data. Therefore, in any optimization problem in which data is transformed to interval-valued probabilities should undergo a preprocessing to obtain the tightest bounds possible associated with the uncertainty data.

2.2 Generalized Uncertainty

Next a review the components of the theory of generalized uncertainty for the ten types of generalized uncertainties of interest is presented. It is shown that these can be translated into the context of IVPMs, The ten types are:

1. Generalized uncertainties directly translating into pairs of functions that bound the uncertainty

   (a) Intervals [5, 6];
   (b) Fuzzy Intervals [7, 8];
   (c) Possibility/Necessity Measures [2, 9];
   (d) Interval-valued Probability [10];
   (e) P-Boxes [11, 12];
   (f) Clouds [13, 14];

2. Generalized uncertainty that bound the uncertainty with a confidence limit—Kolmogorov, Smirnov Statistics [15];
3. Generalized uncertainties that require further hypotheses to translate them into pairs of distributions that bound the uncertainty

   (a) Belief/Plausibility Measures [16];
   (b) Probability Intervals [17];
   (c) Random Sets [18].

The relationship among each of the types of generalized uncertainties is depicted in by Fig. 2.1. As can be seen in Fig. 2.1, generalized uncertainty types either are or have direct translations into interval-valued probabilities. The importance of the relationships as depicted by Fig. 2.1 is that if we have any of our generalized uncertainty types as input data, they can be, in principle, translated into interval-valued probabilities. Thus, to solve optimization under generalized uncertainty for any of these types it suffices to focus on interval-valued probability optimization as we do in Chap. 6. Let us begin by recalling the definition of possibility and necessity measure and possibility and necessity distribution. Then, let us prove that intervals and fuzzy intervals are possibility measures and that fuzzy intervals produce dual possibility and necessity pairs as we have illustrated several times, for example, in Fig. 2.2. Some of the justification of Sects. 2.2.1 and 2.2.2 is found in Sect. 2.2.3.

2.2.1 Intervals

The first type of generalized uncertainty is that of interval. As discussed earlier intervals have a long history as a method for modeling certain types of incomplete information, including rounding error. An interval that represents partial or incomplete information can be modelled as a possibility distribution. Given interval $[a, b]$, $a, b \in \mathbb{R}$, $a \leq b$ where we are using Zadeh's definition of possibilistic distribution [19] (here we use a subscript $[a, b]$ on *pos* to denote that it is the possibility *distribution* of $[a, b]$)

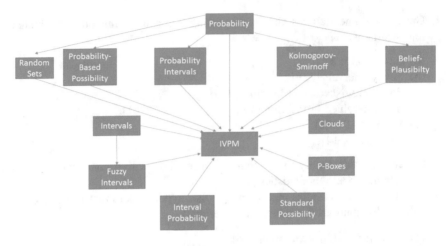

Fig. 2.1 Generalized uncertainty relationships

$$pos_{[a,b]}(x) = \mu_{[a,b]}(x) = \begin{cases} 0 \text{ if } x < a, x > b \\ 1 \text{ if } x \in [a,b] \end{cases}$$

then for any $A \subseteq \mathbb{R}$, $Pos_{[a,b]}(A) = \sup_{x \in A} pos_{[a,b]}(x)$.

So $pos_{[a,b]}$ is simply the characteristic or indicator function. By convention, $Pos_{[a,b]}(\emptyset) = 0$, and if $A = \mathbb{R}$, $Pos_{[a,b]}(\mathbb{R}) = 1$. We can think of $Pos_{[a,b]}$ as modeling some unknown entity y for which all that is known is that it lies in the interval $[a, b]$. Thus $Pos_{[a,b]}(A)$ gives the possibility that $y \in A$.

Now let $A = [a, b]$ and $B = [c, d]$ be two intervals.

$$\text{for any } C \subseteq \mathbb{R}, \ Pos_{A \cup B}(C) = \sup_{x \in C} pos_{A \cup B}(x).$$

where $pos_{A \cup B}(x)$ is the characteristic function for the union of the two intervals. If $x \in A$ or $x \in B$, then the supremum is 1 so that

$$
\begin{aligned}
Pos_{A \cup B}(A \cup B) &= \sup_{x \in A \cup B} pos_{A \cup B}(x) = 1 \\
&= \sup\{Pos_A(A), Pos_B(B)\} \\
&= \sup\{1, 1\} \\
&= 1.
\end{aligned}
$$

Remark 32 The notation Pos_A refers to the possibility **measure**. The notation $pos_A(x)$ refers to the possibility **distribution**. Given a measure, Pos that is defined on singleton subsets, a distribution pos is simply

$$pos(x) = Pos(\{x\}).$$

Given a possibility distribution $pos(x)$, a possibility measure can be defined as

$$Pos(A) = \sup_{x \in A} pos(x).$$

However, there are many ways to construct possibility measures and distributions consistent with their axioms or definition.

2.2.2 Fuzzy Interval

The second type of generalized uncertainty is the fuzzy interval. A fuzzy interval that represents partial or incomplete information is a possibility distribution. For simplicity, let the fuzzy interval $\widetilde{A}$ be a trapezoid $a/b/c/d$, $a < b < c < d \in \mathbb{R}$ so that

$$pos_A(x) = \mu_A(x) = \begin{cases} 0 \text{ if } x < a, x > d \\ \frac{1}{b-a}x - \frac{a}{b-a} \text{ if } a \le x \le b \\ 1 \text{ if } x \in (b, c) \\ -\frac{1}{d-c}x + \frac{d}{d-c} \text{ if } c \le x \le d \end{cases}.$$

We leave the proof as an exercise since it is similar to that which showed that an interval is a possibility.

2.2.3 Possibility/Necessity Measures and Distributions

The third generalized uncertainty type is possibility and necessity. This section briefly develops possibility and necessity theory. A more detailed development can be found in [2, 9]. We limit ourselves to quantitative possibility measures which is a set-valued map

$$Pos(A) : X \subseteq \mathbb{R} \to [0, 1].$$

Any complete ordered lattice could be used instead of [0, 1] but we will use the interval [0, 1].

One of the most general set-valued functions is

$$g : A \subseteq X \to [0, 1], A \in \sigma(X), \tag{2.1}$$

together with a structure of the subsets of X, $\sigma(X)$, a sigma algebra on the set X, from which we pick A. A reasonable set of assumptions on g are the following:

$$g(\emptyset) = 0, \text{ and } g(X) = 1 \tag{2.2}$$
$$A \subseteq B \Rightarrow g(A) \leq g(B). \tag{2.3}$$

The assumptions (2.1)–(2.3) define a *fuzzy measure*, which is a set-valued function. A consequence of (2.2) and (2.3) is:

$$g(A \cap B) \leq \min\{g(A), g(B)\} \tag{2.4}$$
$$g(A \cup B) \geq \max\{g(A), g(B)\}. \tag{2.5}$$

The generalized uncertainty called possibility measure and necessity measure obeys (2.1)–(2.3) where possibility satisfies (2.5) with equality and necessity satisfies (2.4) with equality. We will be interested in possibility measures and their associated distribution. Axioms (2.1)–(2.5) with equality in (2.4), and (2.5), mean that the underlying sets that satisfy these axioms are nested. On the other hand, nested sets of subsets of the sigma algebra of X will generate a possibility measure. These two facts will be proved in the sequel.

Definition 33 (*see for example Chap. 7 of* [2]) A **possibility measure**, axiomatically defined, is a set-valued function Pos over the σ_X of a universal set $X \subset \mathbb{R}$,

$$Pos : \sigma_X \rightarrow [0, 1]$$

such that

$$Pos(\emptyset) = 0, \ Pos(X) = 1, \tag{2.6}$$
$$Pos(A \cup B) = \max\{Pos(A), Pos(B)\}, \ A, B \in \sigma(X) \tag{2.7}$$
$$\text{If } A_i \in \sigma_X, \ \bigcup_{i \in I} A_i \in \sigma_X, \text{ then}$$
$$Pos\left(\bigcup_{i \in I} A_i\right) = \sup_{i \in I} Pos(A_i) \tag{2.8}$$

Example 34 Suppose we have two candidates for one position. Let A be the people who would possibly vote for candidate one and B be the people who would possibly vote for candidate two. The possibility of the election outcome of the votes from people who would vote for candidate one or two $A \cup B$, is given by,

$$Pos(A \cup B) = \max\{Pos(A), Pos(B)\}.$$

Definition 35 Given a *possibility measure, Pos,* its *dual, Nec,* called a **necessity measure** is defined such that for any measurable set $A \in \sigma_X$

$$Nec(A) = 1 - Pos(A^C), \tag{2.9}$$

where A^C is the complement of A which by the definition of a σ_X is in the sigma algebra.

It can be shown that if (2.9) holds,

$$Nec(\emptyset) = 0, \ Nec(X) = 1,$$
$$Nec(A \cap B) = \min\{Nec(A), Nec(B)\}.$$

A *necessity distribution*, associated with the necessity measure Nec is the function

$$nec(x) : X \to [0, 1]$$

where

$$nec(x) = Nec(\{x\}). \tag{2.10}$$

Example 36 Possibility and necessity measures: The transportation department of a city surveys 100 bus riders as to how many people will fit in a bus during rush hour. Forty riders say up to 40 people, thirty people say up to 50, and thirty people say up to 60. In this case

$$U_1 = \{1, ..., 40\}$$
$$U_2 = U_1 \cup \{41, ..., 50\}$$
$$U_3 = U_2 \cup \{51, ..., 60\}$$

so that $U_1 \subset U_2 \subset U_3$. Therefore, we can define a possibility and necessity measure as follows.

$$m(U_1) = 0.40$$
$$m(U_2) = 0.30$$
$$m(U_3) = 0.30$$
$$Pos(U_1) = Pos(U_2) = Pos(U_3) = 1.00$$
$$Pos(\{1\}) = Pos(\{2\}) = ... = Pos(\{40\}) = 1.0$$
$$Pos(\{41\}) = Pos(\{42\}) = ... = Pos(\{50\}) = 0.6$$
$$Pos(\{51\}) = Pos(\{52\}) = ... = Pos(\{60\}) = 0.3$$

For necessity, we have,

$$Nec(U_1) = 1 - Pos(U_1^C) = 0.4$$
$$Nec(U_2) = 1 - Pos(U_2^C) = 0.7$$
$$Nec(U_3) = 1 - Pos(U_3^C) = 1.0$$

Remark 37 Given the definition of possibility measure, Definition 33, we have:
(1) Monotonicity, that is, $A \subseteq B \Rightarrow Pos(A) \leq Pos(B)$;
(2) An underlying nested set of subsets of σ_X.

These two properties are shown subsequently.

Remark 38 We generally do not construct possibility measures from the definition. What we shall see is that there are a variety of ways to construct a possibility measure. To insure that what we construct is in fact a possibility measure, we will need to show that the variety of construction methods satisfy conditions (2.6), (2.7), and (2.8).

Definition 39 Given a possibility measure, *Pos*, a **possibility distribution** associated with *Pos* is a function (on elements of the universal set)

$$pos(x) : X \to [0, 1]$$

such that
$$pos(x) = Pos(\{x\}). \tag{2.11}$$

Note that singletons are part of the σ_X when we use the power set. Since in optimization we will be dealing with existent objects, there is at least one $x \in X$ such that $pos(x) = 1$ though, it is not always known what this value x is. For example, we know in theory, that there exists a minimum radiation dose that will kill a cancer cell. However, what this value is for a particular body, a particular cancer coming from a particular radiation treatment machine, is not known deterministically as a real number. One can obtain a number, but in reality, this number is the best guess based on experience and experimental results. But there exists such a number since there is a radiation level that will kill a cancer tumor cell (and also the patient) and there exists a radiation level that will not kill a cancer tumor cell (zero units of radiation).

The original work by Zadeh [19] on possibility theory started with a possibility distribution which was derived from a fuzzy set. Let $\mu_A(x)$ be a membership function of a given fuzzy set $\tilde{A}$. Then the possibility distribution associated with the fuzzy set $\tilde{A}$, as originally defined by Zadeh, is

$$pos_A(x) = \mu_A(x) \tag{2.12}$$

bearing in mind that the semantics between possibility and fuzzy sets are different.

Remark 40 Zadeh's definition of possibility (2.12) can be, and indeed is, confusing since it appears that there is no difference between fuzzy and possibility.

We address this issue subsequently since fuzzy and possibility are different concepts. However, let us look at an example of Zadeh's definition (2.12).

Example 41 Suppose one is told to meet a "tall" female who will be arriving at the baggage claim of an airport at a particular day/time. "Tallfemale" is a fuzzy set $\mu_{Tallfemale}(x)$ that describes the degree to which a female x is tall. As people come into the baggage claim area, one looks at each female x and obtains a degree of certainty $Tallfemale(x)$. In this case we have $Tallfemale(x) = \mu_{Tallfemale}(x)$. Note that here, $Tallfemale(x)$ is my measure of the partial information about a "tall female" I am to meet at the baggage claim.

The fuzzy set, "tallfemale", $\mu_{Tallfemale}(x)$, is gradualness whereas the possibility distribution $Tallfemale(x)$ is my evaluation of the degree to which a person x at the baggage belongs to the class (the fuzzy set), tall female. That is, $Tallfemale(x)$, is lack of information of what tall is in the context of a female (as opposed to professional basketball players or tall six year old children) and not transitional set belonging. Klir and Yuan [2] state the following:

> A *fuzzy set* F defines the degree to which x belongs to F, not the degree to which evidence supports the fact that x is F. A *possibility measure* is one that assigns a degree of certainty that an element belongs to a set F. It is the degree to which the evidence supports that x is F.

For our example, x is a particular female at the baggage claim and the evaluation of this is my estimation of height of females who are seen, which supports (or not), whether the person is a tall female. The value of this evaluation is $Tallfemale(x)$ is obtained via the fuzzy set $\mu_{tallfemale}(x)$.

One interpretation of possibility is simply as an ordering of a collection of possible outcomes of an uncertain event with $pos(x) > pos(y)$ implying that outcome x is considered more possible than outcome y. In this setting $pos(x)$ is akin to a Demster [16] and Shafer [20] plausibility function. In fact, another way to construct a possibility measure is via plausibility measures over nested sets.

A fuzzy interval such as that depicted in Fig. 2.2, as we mentioned, can be viewed as encoding a family of cumulative probability distributions. Both upper possibility and lower necessity, the magenta and light red distributions of Fig. 2.2 below, are cumulative probability distribution functions. Therefore, given a fuzzy interval as a piece of incomplete information, it generates a family of CDFs bounded by a upper possibility and lower necessity pair as depicted in Fig. 2.2. A possibility description of the value of a parameter (or entity) which encapsulates what is known about the possible values of that parameter or entity, generally uses a (single) fuzzy interval as its distribution function, for example, the trapezoid in Fig. 2.2.

Let us next show that the upper function depicted in Fig. 2.2 is indeed a possibility whereas the lower function, depicted in the same figure, is indeed a necessity. Without loss of generality, we assume that our fuzzy interval is a trapezoidal fuzzy interval.

Proposition 42 *Given a trapezoidal fuzzy interval a/b/c/d, with $a < b < c < d \in \mathbb{R}$ and*

$$pos_{trap}(x) = \begin{cases} 0 \text{ if } x < a, x > d \\ \frac{1}{b-a}x - \frac{a}{b-a} \text{ if } a \leq x \leq b \\ 1 \text{ if } b \leq x \leq c \\ \frac{1}{d-c}x - \frac{c}{d-c} \text{ if } c \leq x \leq d \end{cases}. \tag{2.13}$$

Define the upper function

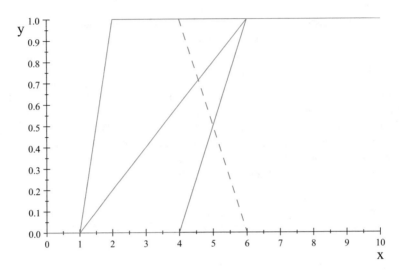

Fig. 2.2 Fuzzy interval—possibility, necessity

$$\overline{pos}_{trap}(x) = \begin{cases} 0 \ if \ x < a \\ \frac{1}{b-a}x - \frac{a}{b-a} \ if \ a \leq x \leq b \\ 1 \ if \ x > b \end{cases} . \qquad (2.14)$$

Both 2.13 and 2.14 are possibility distributions.

Proof The proof is left as an exercise. *Hint,* see the next section and Sect. 2.2.1 or adapt the proof of Proposition 43 to possibility. ∎

Proposition 43 *Given a trapezoidal fuzzy interval a/b/c/d, with $a < b < c < d \in \mathbb{R}$ distribution 2.13, the lower function*

$$nec_{trap}(x) = \begin{cases} 0 \ if \ x < c \\ \frac{1}{d-c}x - \frac{c}{d-c} \ if \ c \leq x \leq d \\ 1 \ if \ x > d \end{cases} \qquad (2.15)$$

is the dual necessity distribution with respect to the possibility Eq. (2.13). Moreover, Eq. (2.15) is also a possibility.

Proof Let fuzzy interval be a/b/c/d be given by (2.13) and (2.14).

$$Nec(A) = 1 - Pos(A^C)$$
$$= 1 - \sup_{x \in A^C} pos_{trap}(x).$$

Let $A = (-\infty, c) \Rightarrow A^C = [c, \infty)$. Therefore

$$Nec(A) = 1 - \sup_{x \in [c, \infty)} pos_{trap}(x)$$

$$= 1 - 1 = 0.$$

This means that

$$nec(x) = 0 \; \forall x \in (-\infty, c).$$

Now let $A = (-\infty, x), x \in (c, d) \Rightarrow A^C = [x, \infty), c < x < d.$

This means that

$$Nec(A) = 1 - \sup_{[x, \infty)} pos_{trap}(x), c < x < d$$

$$= 1 - \left(-\frac{1}{d - c} x + \frac{d}{d - c} \right)$$

$$= 1 + \frac{1}{d - c} x - \frac{d}{d - c}$$

$$= \frac{1}{d - c} x + \frac{d - c}{d - c} - \frac{d}{d - c}$$

$$= \frac{1}{d - c} x - \frac{c}{d - c}.$$

Lastly, let $A = (-\infty, x), x \in [d, \infty) \Rightarrow A^C = [x, \infty), x \geq d.$

This means that

$$Nec(A) = 1 - \sup_{[x, \infty)} pos_{trap}(x), d \leq x$$

$$= 1 - 0$$

$$= 1.$$

Thus, $nec(x) = 1, \forall x \in [d, \infty)$ and the first part of the theorem is proved.

Since $Nec(A)$ is similar to $Pos(A)$ except it is shifted. Therefore, it too is a possibility. ∎

Remark 44 The dual necessity, with respect to the **fuzzy interval** $pos_{trap}(x)$, Eq. (2.13) is also a possibility.

There is a relationship between possibility measures on X and countable nested sets of a sigma-algebra σ_X. In particular, given a possibility measures defined on a countable set of nested sets of a σ_X, it is a possibility measure on the entire σ_X. Conversely, if we have a possibility measure on a σ_X, it is a possibility on any countable set of nested sets of the σ_X. That is, possibility measure can be thought as inducing an underlying generating set of nested subsets. We prove these assertions next.

Theorem 45 *Given a possibility measure, P_Π on any countable set of nested subsets $\Pi = \{A_k, A_k \in \sigma_X, k = 0, 1, 2, ..., \text{ and } X\}$ such that $A_0 = \emptyset$ and $A_i \subseteq A_j, \forall i \leq j$. Then P_Π induces a possibility measure on all of σ_X.*

Proof Let $A \in \sigma_X$ and Π be defined according to the hypothesis. Define

$$Pos(A) = \begin{cases} \sup_{A \cap A_k \neq \emptyset} P_\Pi(A_k), A_k \in \Pi, A \in \sigma_X, \ k = 1, 2, ... \\ 1 \qquad\qquad\qquad\qquad\qquad \text{otherwise} \\ 0 \qquad\qquad\qquad\qquad\qquad \text{for } A = \emptyset \end{cases}.$$

(1) $Pos(\emptyset) = 0$ holds since $\emptyset \cap A_k = \emptyset, \forall k$.

(2) $Pos(X) = 1$ by definition.

(3) Let $A \subseteq B$, for any $A, B \in \sigma_X$. From the definition, it is clear that $Pos(A) \leq Pos(B)$. Thus Pos is a fuzzy measure since it satisfies (2.1)–(2.3). This means that

$$Pos(A \cup B) \geq \max\{Pos(A), Pos(B)\}, \forall A, B \in \sigma_X \qquad (2.16)$$

according to (2.5). It remains to be shown that we have equality in (2.16) for Pos to be a possibility measure, that is, it must satisfy (2.7). However, if A and/or B are either $\emptyset$ or X, we trivially have equality. So we assume that neither A nor B are $\emptyset$ or X. Thus

$$Pos(A \cup B) = \sup_{\{A \cup B\} \cap A_k \neq \emptyset} P_\Pi(A_k), A_k \in \Pi, A, B \in \sigma_X, k > 0 \text{ finite}$$

$$= \sup_{[\{A \cap A_k\} \cup \{B \cap A_k\}] \neq \emptyset} P_\Pi(A_k)$$

$$\leq \max\{\sup_{A \cap A_k \neq \emptyset} P_\Pi(A_k), \sup_{B \cap A_k \neq \emptyset} P_\Pi(A_k)\}$$

$$= \max\{Pos(A), Pos(B)\}. \qquad (2.17)$$

Inequalities (2.16) and (2.17) imply that $Pos(A \cup B) = \max\{Pos(A), Pos(B)\}$. ∎

The converse is also true. That is, given a possibility measure on σ_X, it generates a set of countable set of nested sets.

Theorem 46 *Let $Pos(A)$ be a possibility measure on σ_X. Then, there exists a countable nested set of subsets $A_k, k = 0, 1, 2, ...$ such that $A_i \subseteq A_j, \forall i < j$ with the least subset being the empty set and the maximum set being the universal set X, and a possibility measure Pos_Π on Π where*

$$\Pi = \{A_k, A_k \in \sigma_X, k = 0, 1, 2, ..., \text{ and } X, \text{ with } A_0 \equiv \emptyset\}.$$

Proof Let $I = \{1, 2, ...\}$ and T, a partition of X,

$$T = \{T_i | T_i \subseteq X, i \in I, T_0 = \emptyset\}$$

where $X = \cup_{i \in I} T_i$, and

$$\emptyset = T_i \cap T_j \text{ for } i \neq j, \text{ with } i, j \in I.$$

At least one such partition of X in disjoint subsets exists, $T = \{\emptyset, X\}$, which is the trivial partition. Let

$$A_k = \cup_{i=1}^{k} T_i$$
$$A_0 = \emptyset = T_0,$$
$$\Pi = \{A_0, A_1, ...\},$$

and define

$$Pos_\Pi(A_k) = \max_{0 \leq i \leq k} Pos(T_i).$$

Clearly, Π is a nested set of subsets of σ_X and Pos_Π is a possibility measure on Π. ∎

Remark 47 The significance of these two theorems, as was mentioned, is that underlying possibility measures are these nested subsets that generate the measures and possibility measures generate these nested set of subsets. Moreover, if we are to construct possibility necessity pairs, we either need to begin with a set of nested subsets, or if we construct a distribution pair, there is nestedness that underlies the sets over which the pair of distribution measures operate.

We now list some simple standard properties associated with possibility and necessity measures (see [2]).

1. $Nec(A) + Nec(A^C) \leq 1$
2. $Pos(A) + Pos(A^C) \geq 1$
3. $Nec(A) + Pos(A^C) = 1$
4. $\min\{Nec(A), Nec(A^C) = 0$
5. $\max\{Pos(A), Pos(A^C) = 1$
6. $Nec(A) > 0 \Rightarrow Pos(A) = 1$
7. $Pos(A) < 1 \Rightarrow Nec(A) = 0.$

2.2.4 Interval-Valued Probability

The fourth generalized uncertainty type of interest is the interval-valued probability. What is presented next is an abbreviated version of [21–23]. A basis for linking various methods of uncertainty representation is examined next. One reason for studying the various generalized uncertainties and their interrelationships is that when we have a mixture of uncertainties occurring in a single constraint equation/inequality, all uncertainties need to be translated into the most general uncertainty which has as particular cases the set of uncertainties that appear in the constraint. This is true of IVPMs since they are the most general of our ten generalized uncertainty types. Once the various uncertainties are translated into the over arching one, upper and lower distribution functions bounding the set of uncertainties are constructed/computed. IVPM is a foundation of our approach. This section begins by defining what is meant by an IVPM. The set of functions defined by IVPM is a representation of the uncertainty associated with the given information about an unknown probability measure. Throughout, arithmetic operations involving set functions are in terms of (constraint) interval functions [? [5, 24]] or intervals, which are in terms of (constraint) interval arithmetic [5, 25]. $Int_{[0,1]} \equiv \{[a, b] \mid 0 \leq a \leq b \leq 1\}$ denotes the set of all interval contained in $[0, 1]$.

We first define, in a formal way, what is called an interval-valued probability measure as used by Weichselberger [10]. Weichselberger's definition begins with a set of probability measures, not necessarily the tightest, called $R - probabilities$. Then he defines an interval probability as a set function providing lower and upper bounds on the probabilities calculated from these measures. $F - probabilities$, as defined by Weichselberger, are simply the tightest bounds possible of all $R - probabilities$. This definition is followed by a demonstration that various forms of uncertainty representations (interval, probability, possibility, cloud, P-Boxes, and probability interval) all which can be represented by IVPMs. Then we show how interval-valued probability measures can be constructed from lower and upper bounding cumulative distribution functions. This is followed by an extension principle for a function of uncertain variables represented by interval-valued probability measures and integration with respect to interval-valued probability measures. Both of these definitions will be useful in analyzing problems involving uncertainty represented by interval-valued probability measures.

Note that generalized expected values require integration. This means, when we translate uncertainty types in IVPMs, we need an extension principle for IVPMs in order to do mathematical analysis, optimization in our case, with these types of uncertainties. This will be done subsequently.

Throughout this section we will be primarily interested in interval-valued probability defined on the Borel sets, denoted $\mathcal{B}$, on the real line and real-valued random variables. Recall that Borel sets on $\mathbb{R}$ is its $\sigma-$field. The basic definitions from Weichselberger (with slight variation in notation) are presented next.

Definition 48 (*Weichselberger* [10]) Given measurable space $(S, \mathcal{A})$, an interval valued function $i_m : \mathcal{A} \to Int_{[0,1]}$ is called an **R-probability** measure if:
(a) $i_m (A) = \left[a^- (A), a^+ (A)\right] \subseteq [0, 1]$ with $a^- (A) \leq a^+ (A)$
(b) There exists a probability measure Pr on $\mathcal{A}$ such that

$$\forall A \in \mathcal{A}, \Pr (A) \in i_m (A)$$

By an **R-probability field** we mean the triple $(S, \mathcal{A}, i_m)$.

Definition 49 (*Weichselberger* [10]) Given an R-probability field $\mathcal{R} = (S, \mathcal{A}, i_m)$ the set

$$\mathcal{M}(\mathcal{R}) = \{\Pr \mid \Pr \text{ is a probability measure on } \mathcal{A} \text{ such that } \forall A \in \mathcal{A}, \Pr(A) \in i_m(A)\}$$

is called the **structure** of $\mathcal{R}$.

Definition 50 (*Weichselberger* [10]) An R-probability field $\mathcal{R} = (S, \mathcal{A}, i_m)$ is called an **F-probability field**, if $\forall A \in \mathcal{A}$:
(a) $a^+(A) = \sup\{\Pr(A) \mid \Pr \in \mathcal{M}(\mathcal{R})\}$,
(b) $a^-(A) = \inf\{\Pr(A) \mid \Pr \in \mathcal{M}(\mathcal{R})\}$.

Remark 51 Given a measurable space $(S, \mathcal{A})$ and a set of probability measures P, defining

$$a^+(A) = \sup\{\Pr(A) \mid \Pr \in P\}$$

and

$$a^-(A) = \inf\{\Pr(A) \mid \Pr \in P\}$$

gives an $F - probability$, where P is a subset of the structure. What this means is that we can begin with an arbitrary set of probabilities, where our unknown one is somewhere in this set, and construct an *F-probability field*.

2.2.5 P-Boxes

Our fifth type of generalized uncertainty are P-Boxes. P-Boxes are a generalized uncertainty type since they are pairs of distributions enclosing unknown distributions. Two original publications associated with P-Boxes are [11, 12]. Let a probability density distribution be denoted p. Recall that the cumulative distribution associated with respect to p is

$$F_p(x) = \int_{-\infty}^{x} p(t)dt.$$

Definition 52 A **P-Box** is defined by a pair of cumulative distributions $\underline{F}(x) \leq \overline{F}(x)$, $x \in \mathbb{R}$ and is the set of cumulative distribution functions $F_p(x)$ such that

$$\underline{F}(x) \leq F_p(x) \leq \overline{F}(x).$$

Given a P-Box, it is clear that

$$PB = \{F_p(x) \mid \underline{F}(x) \leq F_p(x) \leq \overline{F}(x)\} \neq \emptyset$$

since both $\underline{F}(x)$ and $\overline{F}(x)$ are cumulatives that belong to PB. Therefore, we can take the bounding cumulative functions $\underline{F}(x)$ and $\overline{F}(x)$ as our possibility pairs since cumulative distributions are possibilities (see Propositions 42 and 43). Moreover, the bounding functions also define the extremes of an interval-valued probability measure as we will see. A more extensive presentation of P-Boxes is [11]. The problem, of course, is how to construct PB when confronted with uncertain information where we wish PB to be minimal in some sense. This construction of distributions is left to Chap. 3.

2.2.6   Cloud

Our sixth generalized uncertainty type is a cloud. The idea of a cloud is to enclose uncertainty in such a way that we have probabilistic-like characteristics without being given or knowing upper/lower cumulative distributions but we know upper an lower functions with some specific properties delineated below. The set of random variables belonging to a cloud is non-empty, so that clouds are well-defined and within the theory of probability. That is, given any cloud, there exists a random variable belonging to the cloud (see [13]). The original impetus was to be able to model analytically in the presence of missing information, missing precision of concepts, and/or dealing with imprecise measurements and there is a well-defined semantic to this lack of information.

Definition 53 ([13]) A **cloud** over a set M is a mapping $\mathbf{x}$ that associates with each $\xi \in M$ a non-empty, closed, and bounded interval $\mathbf{x}(\xi)$, such that,

$$(0, 1) \subseteq \bigcup_{\xi \in M} \mathbf{x}(\xi) \subseteq [0, 1]. \tag{2.18}$$

$\mathbf{x}(\xi) = [\underline{\mathbf{x}}(\xi), \overline{\mathbf{x}}(\xi)]$, is called the **level** of ξ in the cloud $\mathbf{x}$ where $\underline{\mathbf{x}}(\xi)$, and $\overline{\mathbf{x}}(\xi)$ are the **lower** and **upper level**, respectively, and $\overline{\mathbf{x}}(\xi) - \underline{\mathbf{x}}(\xi)$ is called the **width** of ξ. When the width is zero for all ξ, the cloud is called a **thin cloud**.

Remark 54 $\mathbf{x}(\xi)$, as can be seen in Fig. 2.3, is a **vertical** interval for each ξ, that is, an interval with respect to the y-axis. Of course the width is the vertical width.

When doing analysis over the domain of real numbers, one needs algebraic operations over clouds in addition to concepts of integration, differentiation, measure, limits.

Definition 55 ([13]) A real **cloudy number** is a cloud over the set $\mathbb{R}$ of real numbers. $\chi_{[a,b]}$ (χ being the characteristic function) is the cloud equivalent to an interval $[a, b]$ which is

$$\underline{\mathbf{x}}(\xi) = 0$$
$$\overline{\mathbf{x}}(\xi) = \begin{cases} 1 \text{ if } \xi \in [a, b] \\ 0 \text{ otherwise} \end{cases}.$$

A **cloudy vector** is cloud over $\mathbb{R}^n$, where each component is a cloudy number.

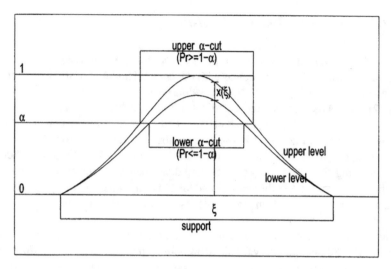

Fig. 2.3 Cloud

An interval provides only information about the support of the probability density
function without additional probabilistic content. Neumaier [13] states that depen-
dence or correlation between uncertain numbers (or the lack thereof) can be modeled
by considering them jointly as components of a cloudy vector. Moreover, in many
applications the level $\mathbf{x}(\xi)$ may be interpreted as giving lower and upper bounds
on the *degree of suitability* of $\xi \in M$ as a possible scenario for data modeled by
the cloud $\mathbf{x}$. This degree of suitability can be given a probability interpretation by
relating clouds to random variables (see Eq. (2.19) below). We say that a random
variable $\mathbf{x}$ with values in M belongs to a cloud $\mathbf{x}$ over M, and write $x \in \mathbf{x}$, if

$$\Pr(\underline{x}(x) \geq \alpha) \leq 1 - \alpha \leq \Pr(\overline{x}(x) > \alpha) \ \forall \, \alpha \in [0, 1]. \qquad (2.19)$$

Since Pr denotes the probability measure, its argument, the sets consisting of all ξ
$\in M$ where $\underline{x}(x) \geq \alpha$ and $\overline{x}(x) > \alpha$ are required to be measurable in σ_M the sigma
algebra on M consisting of all sets $A \subseteq M$ for which $\Pr(x \in A)$ is defined. This
approach gives clouds an underlying interpretation as the class of random variables
x with $x \in \mathbf{x}$.

Given a fuzzy set, a fuzzy cloud is a cloud where the fuzzy membership function
is the upper level and the x-axis is the lower level. The interpretation of a fuzzy set
being bounded by the membership function and the x-axis was first advocated by
Dubois, Moral, and Prade [26]. Since there exists at least one random variable in
every cloud [13], this interpretation is meaningful. Moreover, it is in this way that a
fuzzy set may be associated with uncertainty. However, we do not pursue this further.

2.2.7 Kolmogorov-Smirnov Statistic

A seventh generalized uncertainty type is the Smirnov-Kolmogorov statistic generated distributions. Many reference with cogent explanations of the Kolmogorov-Smirnov Statistic (KSS) can be found. *Matlab* has a KSS in its statistical toolbox (kstest2) as does the open source statistical software package *R*. This is an approach that uses confidence limits to construct upper and lower distributions on uncertainty. That is, the KSS, in the way it is used for enclosing uncertainty, forms confidence bands of an unknown cumulative distribution function $F(x)$. Finding a confidence band enclosing $F(x)$, begins by finding confidence bands around each x. Chapter 3 will show how to construct these upper and lower bounds on the underlying unknown cumulative distribution, given (incompletely specified) statistics that describe the problem.

The next three uncertainty types are those that require special properties in order that to translate them into interval-valued probability measure or possibility/necessity measures. These are discussed next but not developed beyond what we present here.

2.2.8 Dempster-Shafer Belief/Plausibility Measures and Distributions

A eighth generalized uncertainty type is the Dempster-Shafer belief and plausibility measures which was developed by Dempster [16] and Shafer [20, 27].

Definition 56 A belief measure is a set-valued function

$$Bel : \sigma_X \rightarrow [0, 1]$$

such that

$$Bel(\emptyset) = 0$$
$$Bel(X) = 1$$
$$A_k \in \sigma_X, k = 1, ..., K$$
$$Bel(\cup_{k=1}^{K} A_k) \geq \sum_{k=1}^{K} Bel(A_k) - \sum_{j<k} Bel(A_j \cap A_k)$$
$$+ ... + (-1)^{K+1} Bel(\cap_{k=1}^{K} A_k).$$

The last property is called super-additivity. Note that if the A_k are mutually disjoint, then

$$Bel(\cup_{k=1}^{K} A_k) \geq \sum_{k=1}^{K} Bel(A_k)$$

which means belief measures are, in general, distinct from probability. A plausibility measure is a set-valued function

$$Pl : \sigma_X \to [0, 1]$$

such that

$$Pl(\emptyset) = 0$$
$$Pl(X) = 1$$
$$A_k \in \sigma_X, k = 1, ..., K$$
$$Pl(\cup_{k=1}^{K} A_k) \le \sum_{k=1}^{K} Pl(A_k) - \sum_{j<k} P(A_j \cap A_k)$$
$$+ ... + (-1)^{K+1} Pl(\cap_{k=1}^{K} A_k).$$

The last property is called sub-additivity. Note that if the A_k are mutually disjoint, then

$$Pl(\cup_{k=1}^{K} A_k) \le \sum_{k=1}^{K} Pl(A_k)$$

which means that plausibility measures are, in general, distinct from probability theory. Moreover, Bel and Pl are dual, that is,

$$Pl(A) = 1 - Bel(A^C).$$

If σ_X consists of nested sets, then

$$Bel(A \cup B) = \max\{Bel(A), Bel(B)\}$$

and

$$Pl(A \cap B) = \min\{Pl(A), Pl(B)\}.$$

This means that they satisfy the axioms of possibility and necessity.

Note that in the case of a system of nested sets, Bel and Pl are Pos and Nec respectively (see [2]). Some properties of belief and plausibility are:

1. $Bel(A) + Bel(A^C) \le Bel(A \cup A^C) = Bel(X) = 1$;
2. $Pl(A) + Pl(A^C) \ge Pl(A \cup A^C) = Pl(X) = 1$;
3. $Bel(A) \le Pl(A)$.

The construction of belief and plausibility from data usually uses what is called the assignment function which is Definition 65 given below as part of the discussion on random sets.

Example 57 Belief and Plausibility Measures: A poll is taken of a group in a city of 10,000 people who will vote for five mayoral candidates {a, b, c, d, e}. Suppose there are two political parties who have candidates running for mayor, Candidates {a, b} belong to Party A and candidates {c, d, e} belong to Party B. Suppose 3,500 are loyal Party A members who prefer {a,b} but as of yet, have not decided for which of the two they will vote. Likewise, suppose 4,500 are from Party B and will vote for {c, d, e} but as of yet have not decided which of the three candidates they prefer. The remaining 2,000 are undecided. The focal elements are $\{a, b\}$, $\{c, d, e\}$, $\{a, b, c, d, e\}$ where we have as an assignment function (see Definition 65),

$$m(\{a, b\}) = 0.35$$
$$m(\{c, d, e\}) = 0.45$$
$$m(\{a, b, c, d, e\}) = 0.20$$

Note that the assignment function is not monotonic. The assignment function for a set A, m(A), is solely a property of the set A and no other information impacts the value and is the amount of probability left over from the elements of A that are not yet assigned due to lack of information.

$$Bel(\{a, b\}) = \sum_{B \subseteq \{a,b\}} m(B) = 0.35$$
$$Bel(\{c, d, e\}) = \sum_{B \subseteq \{c,d,e\}} m(B) = 0.45$$
$$Pl(\{a, b\}) = 1 - Bel(\{a, b\}^C)$$
$$= 1 - Bel(\{c, d, e\})$$
$$= \sum_{B \cap \{a,b\} \neq \emptyset} m(B)$$
$$= 0.55$$
$$Pl(\{c, d, e\}) = 1 - Bel(\{c, d, e\}^C)$$
$$= 1 - Bel(\{a, b\})$$
$$= \sum_{B \cap \{c,d,e\} \neq \emptyset} m(B)$$
$$= 0.65$$

Note that

$$Bel(\{X\}) = 1 \geq Bel(\{a, b\}) + Bel(\{c, d, e\}) = 0.80$$
$$Pl(X) = 1 \leq Pl(\{a, b\}) + Pl(\{c, d, e\}) = 1.20$$

Example 58 Belief and plausibility measures, assignment function: Given the example of the voters above, suppose of the 2,000 undecided votes, 500 will vote for candidate {a} and another 500 will vote for {b} or {d}. Thus, our sets are

$$A_1 = \{a, b\}, A_2 = \{c, d, e\}, A_3 = \{a\}, A_4 = \{b, d\}, X = \{a, b, c, d, e\}.$$

Our update yields new assignment functions

$$m(A_1) = 0.35, m(A_2) = 0.45, m(A_3) = 0.05, m(A_4) = 0.05, m(X) = 0.10$$

and in terms of belief and plausibility

$$Bel(A_3) = m(A_3) = 0.05,$$
$$Bel(A_1) = m(A_1) + m(A_3) = 0.40$$
$$Pl(A_1) = m(A_1) + m(A_3) + m(X) = 0.50$$

2.2.9 Probability Interval

The ninth uncertainty type is probability interval. Probability intervals provide a way for discrete probability data to be enclosed by bounding distributions. The probability interval is an entity that was developed by [17] and is formally defined as follows.

Definition 59 ([17]) Given a discrete set $X = \{x_1, x_2, ..., x_K\}$, a **probability interval** is the set

$$PI = \{[a_k, b_k], k = 1, ..., K, 0 \le a_k \le b_k \le 1\}$$

such that, for the discrete case,

$$M_{PI}^d = \{Pr \mid a_k \le Pr(\{x_k\}) \le b_k, \ k = 1, ..., K\}.$$

For the continuous case

$$M_{PI}^c = \{F_p(x) \mid a(x) \le Pr((-\infty, x]) \le b(x), \ k = 1, ..., K, 0 \le a(x) \le b(x)\},$$

where $b(x) \le 1$.

Theorem 60 ([28, 29]) *For the discrete case, $M_{PI}^d \ne \emptyset$ if and only if $\sum_{k=1}^{K} a_k \le 1 \le \sum_{k=1}^{K} b_k$.*

Proof It is clear from the properties of probability that if a probability density where to be defined on the interval $[a_k, b_k]$, then, M_{PI}^d would be empty when either $\sum_{k=1}^{K} a_k > 1$ or $\sum_{k=1}^{K} b_k < 1$. However, if

$$\sum_{k=1}^{K} a_k \le 1 \le \sum_{k=1}^{K} b_k,$$

then a probability $p \in M_{PI}^d$ can be constructed as follows.

$$\sum_{k=1}^{K} a_k = \sum_{k=1}^{K-1} a_k + a_K \leq 1.$$

Let $\epsilon_K \geq 0$ be such that

$$\sum_{k=1}^{K-1} a_k + a_K + \epsilon_K = 1,$$

Then $p(x_k) = a_k$, where $p(x_K) = a_K + \epsilon_K$ is a probability such that $p \in M_{PI}^d$. ∎

Theorem 61 *For the continuous case,* $M_{PI}^c \neq \emptyset$ *if and only if* $\int_{-\infty}^{\infty} a(x)dx \leq 1 \leq \int_{-\infty}^{\infty} b(x)dx.$

Proof The proof is similar to Theorem 60.   ∎

Definition 62 ([28, 29]) A discrete probability interval

$$PI = \{[a_k, b_k], k = 1, ..., K, 0 \leq a_k \leq b_k \leq 1\}$$

is **reachable** if there exists $\underline{\text{Pr}}_k, \overline{\text{Pr}}^k \in M_{PI}^d$ such $a_k = \underline{\text{Pr}}_k(x_k)$ and $b_k = \overline{\text{Pr}}^k(x_k)$ for each $k = 1, 2, ..., K$.

Theorem 63 ([28, 29]) *A probability interval* $PI = \{[a_k, b_k], k = 1, ..., K, 0 \leq a_k \leq b_k \leq 1\}$ *is reachable if and only if*

$$\sum_{j \neq k} a_j + b_k \leq 1, k = 1, ..., K$$

and

$$\sum_{j \neq k} b_j + a_k \leq 1, k = 1, ..., K.$$

Proof See [28, 29] for a proof.   ∎

De Campos, Huete, and Moral [17] showed that by defining

$$l(A) = \min_{\text{Pr} \in M_{PI}} P(A) \text{ and } u(A) = \max_{\text{Pr} \in M_{PI}} P(A), \forall A \subseteq X$$

there is a formulation of $l(A)$ and $u(A)$ for the reachability case according to the following theorem, which is the key to translating probability intervals into other types of uncertainty.

Theorem 64 ([28, 29]) *If $PI = \{[a_k, b_k], k = 1, ..., K, 0 \le a_k \le b_k \le 1\}$ is reachable, then (1) $l(x_k) = a_k$ and $u(x_k) = b_k$, (2) $\forall A \subseteq X$,*

$$l(A) = \max \left\{ \sum_{x_k \in A} a_k, 1 - \sum_{x_k \in A^C} b_k \right\}$$

and

$$u(A) = \max \left\{ \sum_{x_k \in A} u_k, 1 - \sum_{x_k \in A^C} l_k \right\}.$$

Proof See [28, 29].   ∎

The significance of the above is that under the conditions of the theorems, upper and lower distributions can be constructed, which in fact can be identified with interval-value probability measures and possibility/necessity measures.

2.2.10 Random Sets

The tenth and last generalized uncertainty type is random set. Random sets comprise the last uncertainty type of interest, because they are quite general. A basic text on random sets is [18]. Notice that from the point of view of uncertainty types as depicted in Fig. 2.1, random sets constructed from belief and plausibility are IVPMs. The name arises from the problem of selecting a sub-sample (a subset) from an entire population at random. However, it has a formal development of its own. Where it comes into play in uncertainty is when we only have probability information on non-singleton subsets rather than probabilities on the elements of subsets. In this case, a random set is generated and from random sets, we are able to create upper and lower bounds which turn out to be belief and plausibilities. In fact, random sets and belief and plausibility measures encode the same uncertainty.

Definition 65 A mapping

$$m : \sigma_X \to [0, 1]$$

such that

$$\sum_{A \in \mathcal{F}} m(A) = 1$$

generates a **random set** $(\mathcal{F}, m)$, where $\mathcal{F} = \{A \in \sigma_X \mid m(A) > 0\}$, and A is called a focal element. The mapping m is called the **basic probability assignment function** or just assignment function.

Example 66 Basic Assignment Function: The assignment function m(A) expresses the proportion to which all available and relevant information support the claim that a particular element $x \in X$ whose characterization in terms of relevant attributes is incomplete belongs to a set A. That is, it is like a possibility in the sense of the textbook by Klir and Yuan. With respect to our Example 57 above, $x \in \{a, b, c, d, e\} = X$ where

$$m\{a, b\} = 0.35,$$
$$m\{c, d, e\} = 0.45,$$
$$m\{a, b, c, d, e\} = 0.20,$$

that is, $m(X) = m\{a, b, c, d, e\} = 0.20$, is the value of unassigned probability since the sets for which there is information are just $\{a,b\}$, $\{c,d,e\}$ and $\{a,b,c,d,e\}$ and $\emptyset$. The focal elements are just the three. Notice that the assignment function is not monotonic.

Given complete information, every element of a finite universe X is assigned so for all $x \in X, m\{x\} = \Pr(x)$ and $\sum_x m\{x\} = 1$. In this case for any set A that is not a singleton including X, $m(A) = 0$. Thus a basic assignment function has the following properties:

1. $A \subseteq B$, it is not required that $m(A) \leq m(A)$ while in probability (and possibility) $\Pr(A) \leq \Pr(B)$
2. It is not required that $m(X) = 1$ while in probability and possibility $\Pr(X) = 1$.
3. There is no dual relationship between $m(A)$ and $m(A^C)$ whereas in probability $\Pr(A) + \Pr(A^C) = 1$.

The basic assignment function is an abstract concept useful in cases in which the precise probability for all sets of the power set of X, $P(X)$, is not known.

Example 67 (**Random set**) For the first mayoral election Example 57, there is the family of subset

$$\mathcal{F} = \{A_1 = \{a, b\}, A_2 = \{c, d, e\}, X = \{a, b, c, d, e\}\},$$

with

$$m(A_1) = 0.35$$
$$m(A_2) = 0.45$$
$$m(X) = 0.20$$

Then $(\mathcal{F}, m)$ is the random set. If a different group of 10,000 in the survey, a different random set will occur, since the $\mathcal{F}$ and m will change. If 10,000 people are added to the already surveyed 10,000 people, the sample space X will increase so that the $\mathcal{F}$ and m will undoubtedly change. For example, if more details about the 4,500 who support Party B is received, an update to the random set will be necessary. That is, suppose of the 4,500 who will vote for someone in Party B, 2,000 will vote for candidate $\{c\}$ while 2,500 are still undecided for which of $\{c,d,e\}$ they will vote, then

$$\mathcal{F}_{new} = \{A_1, A_2, A_3 = \{c\}\}$$

We then have

$$m(A_1) = 0.35$$
$$m(A_2) = 0.45$$
$$m(A_3) = 0.20$$
$$m(X) = 0.20$$

in terms of belief and plausibility measures.

2.3 Equivalences Between Generalized Uncertainty Types

Next how various of our uncertainty types are equivalent will be shown. We begin by showing how intervals, possibility distributions, clouds and (of course) probability measures can define IVPMs on the Borel sets of the real line. That is, our uncertainty types can be transformed into IVPMs.

Remark 68 (An interval defines an IVPM): Let $I = [a, b]$ be a non-empty interval on the real line. On the Borel sets define

$$a^+(A) = \begin{cases} 1 \text{ if } I \cap A \neq \emptyset \\ 0 \quad \text{otherwise} \end{cases},$$

and

$$a^-(A) = \begin{cases} 1 \text{ if } I \subseteq A \\ 0 \text{ otherwise} \end{cases},$$

then

$$i_m(A) = \left[a^-(A), a^+(A)\right]$$

defines an interval-valued probability measure on $(\mathcal{R}, \mathcal{B}, i_m)$. To show that this is an IVPM, first note that if

$$i_m(A) = [0, 0]$$

then

$$\forall k, i_m(A_k) = [0, 0]$$

and for at least one j,

$$i_m(B_j) = [1, 1]$$

or [0, 1] and if

$$i_m\left(B_j\right) = [1, 1]$$

then

$$\forall l \neq k, i_m\left(B_j\right) = [0, 0].$$

Thus

$$\Sigma_{k \in K} A_k^l = \Sigma_{k \in K} A_k^u = 0 \ 1 - \Sigma_{j \in J} B_j^u \leq 0, \ \text{and} \ 1 - \Sigma_{j \in J} B_j^l \geq 0,$$

which shows that

$$\max\left\{1 - \Sigma_{j \in J} B_j^u, \Sigma_{k \in K} A_k^l\right\} = 0 \ \text{and} \ \min\left\{1 - \Sigma_{j \in J} B_j^l, \Sigma_{k \in K} A_k^u\right\} = 0.$$

The proof for the other two cases is similar.

Remark 69 (A probability measure is an IVPM) Let Pr be a probability measure over $(S, \mathcal{A})$. Define $i_m(A) = [\Pr(A), \Pr(A)]$ which is equivalent to having total knowledge about a probability distribution over S. It is easy to see that this meets the definition of an interval-probability measure by noting for example that $\Pr_X(\cup_{k \in K} A_k) = \Sigma_{k \in K} A_k^l = 1 - \Pr_X(\cup_{j \in J} B_j) = 1 - 1 - \Sigma_{j \in J} B_j^u$. Thus modeling with IVPMs includes problems that also include known probability distributions.

Remark 70 (A possibility distribution defines an IVPM) Let

$$pos(x) : x \in S \rightarrow [0, 1]$$

be a regular possibility distribution function and let Pos be the associated possibility measure and Nec the dual necessity measure. Then define

$$i_m(A) = [Nec(A), Pos(A)].$$

Moreover, if a second possibility distribution, $\underline{p}(x) = 1 \ \forall x$ is defined, then the pair $p, \underline{p}$ define a cloud for which $i_m(A)$ is its associated R-probability according to the previous example. Since a cloud exists, there is at least one random variable in it, according to [13]. Therefore $i_m(A)$ is not empty and so well-defined.

The concept of a cloud [13], which was already introduced, can be defined in the context of possibility distributions making its relationship to possibility clearer. Let us then use a slightly different notation for cloud where it will turn out that the upper level is a possibility and the lower level is a necessity though they will not necessarily be duals of each other. That is given an upper level, the lower level is not necessarily its dual but it will be a necessity. Thus, we use $\overline{p}$ to denote the upper level and $\underline{n}$ to denote the lower level of a cloud. The notation for the duals of $\overline{p}$ and $\underline{n}$ are $\overline{n}$ and $\underline{p}$ respectively. Generally, $\overline{p}$ and $\underline{p}$ are distinct functions as are $\overline{n}$ and $\underline{n}$.

Remark 71 From the point of view of interval-valued probabilities, recall that a cloud over set S is a mapping c such that:

(1) $\forall s \in S$, $c(s) = [\underline{n}(s), \bar{p}(s)]$ with $0 \leq \underline{n}(s) \leq \bar{p}(s) \leq 1$,

(2) $(0, 1) \subseteq \bigcup_{s \in S} c(s) \subseteq [0, 1]$.

In addition, random variable X taking values in S is said to belong to cloud c (written $X \in c$) iff

(3) $\forall \alpha \in [0, 1]$, $\Pr(\underline{n}(X) \geq \alpha) \leq 1 - \alpha \leq \Pr(\bar{p}(X) > \alpha)$.

Clouds are closely related to possibility theory. A function $p : S \to [0, 1]$ is called a regular (normalized) possibility distribution function if

$$\sup\{p(x) \mid x \in S\} = 1.$$

By convention, define $\sup\{p(x) \mid x \in \emptyset\} = 0$. Define another distribution function n by

$$n : S \to [0, 1]$$
$$n(x) = 1 - p(x),$$

where by convention $\inf\{n(x) \mid x \in \emptyset\} = 1$.

Now, the concept of a cloud can be stated in terms of certain pairs of consistent possibility distributions (there is at least one domain value whose possibility is equal to 1), which is shown in the following proposition.

Proposition 72 *Let $\bar{p}$ and $\underline{p}$ be a pair of regular possibility distribution functions over set S such that $\forall s \in S$, $\bar{p}(s) + \underline{p}(s) \geq 1$. Then the mapping $c(s) = [\underline{n}(s), \bar{p}(s)]$ where $\underline{n}(s) = 1 - \underline{p}(s)$ is a cloud. In addition, if X is a random variable taking values in S and the possibility measures associated with $\bar{p}, \underline{p}$ are consistent with X then X belongs to cloud c. Conversely, every cloud defines such a pair of possibility distribution functions and their associated possibility measures are consistent with every random variable belonging to the cloud c.*

Proof $\Rightarrow$

(1) Let $\bar{p}, \underline{p} : S \to [0, 1]$ and

$$\bar{p}(s) + \underline{p}(s) \geq 1 \Rightarrow$$
$$\overline{p}(s) + (1 - \underline{n}(s)) \geq 1 \Rightarrow$$
$$\overline{p}(s) \geq \underline{n}(s).$$

This implies that property (1) of Remark 71 is satisfied.

(2) Since all regular possibility distributions satisfy

$$\sup \{p\,(s) \mid s \in S\} = 1 \text{ means that}$$
$$p(s) \leq 1.$$
$$p : S \to [0, 1] \text{ then}$$
$$\inf \{p\,(s) \mid s \in S\} \geq 0 \text{ means that}$$
$$p(s) \geq 0.$$

This means that property (2) of Remark 71 holds. Therefore c is a cloud. Now assume consistency. Then

$$\alpha \geq Pos\,\{s \mid \overline{p}\,(s) \leq \alpha\}$$
$$\geq \Pr\{s \mid \overline{p}\,(s) \leq \alpha\}$$
$$= 1 - \Pr\{s \mid \overline{p}\,(s) > \alpha\} \text{ or}$$
$$\Pr\{s \mid \overline{p}\,(s) > \alpha\} \geq 1 - \alpha,$$

which gives the right-hand side of the required inequalities.

$$1 - \alpha \geq Pos\,\left\{s \mid \underline{p}\,(s) \leq 1 - \alpha\right\}$$
$$\geq \Pr\left\{s \mid \underline{p}\,(s) \leq 1 - \alpha\right\}$$
$$= \Pr\left\{s \mid 1 - \underline{p}\,(s) \geq \alpha\right\}$$
$$= \Pr\left\{s \mid \underline{n}\,(s) \geq \alpha\right\}$$

which gives the left-hand side of consistency.

⇐The converse was proved in Sect. 5 of [13].   ∎

Remark 73 (A cloud defines an R-probability field) Let c be a cloud over the real line. Let Pos^1, Nec^1, Pos^2, Nec^2 be the possibility measures and their dual necessity measures relating to $\overline{p}\,(s)$ and $\underline{n}\,(s)$ where $\overline{p}\,(s)$ is the upper level and $\underline{n}\,(s)$ is the lower level as in Remark 71. That is,

$$Pos_1(A) = \sup_{a \in A} \overline{p}(a)$$
$$Pos_2(A) = \sup_{a \in A} \underline{n}(a)$$
$$Nec_1(A) = 1 - Pos_1(A^C)$$
$$Nec_2(A) = 1 - Pos_2(A^C).$$

Define

$$i_m\,(A) = [\max\{Nec_1\,(A)\,, Nec_2\,(A)\}\,, \min\{Pos_1\,(A)\,, Pos_2\,(A)\}]\,.$$

In [13] Neumaier proved that every cloud contains a random variable X. Consistency requires that $\Pr (X \in A) \in i_m (A)$. Thus $i_m (A)$ contains at least one random variable so that the measure is non-empty. This means it is an R-probability field.

Remark 74 (A possibility distribution defines an R-probability field) Let $p : S \to [0, 1]$ be a regular possibility distribution function and let Pos be the associated possibility measure. Then

$$i_m (A) = [0, Pos (A)].$$

Moreover, if we define a second possibility distribution, $\underline{p} (x) = 1 \ \forall x$ then the pair $p, \underline{p}$ define a cloud for which $i_m (A)$ is its associated R-probability according to the previous example. Since we have a cloud, there exists at least one random variable in the cloud according to [13] and so $i_m(A)$ is not empty so well-defined.

Remark 75 (A probability interval defines an R-probability field) Let

$$M_{PI}^d = \{\Pr \mid a_k \leq \Pr(\{x_k\}) \leq b_k, \ k = 1, ..., K\} \neq \emptyset.$$

It is clear that if we pick $\underline{p}_k = a_k$ and $\overline{p}_k = b_k$ we have a discrete R-probability. For the continuous case we have,

$$M_{PI}^c = \{F_p(x) \mid a(x) \leq \Pr((-\infty, x]) \leq b(x), \ k = 1, ..., K, 0 \leq a(x) \leq b(x)\}.$$

Choosing $\underline{p}(x) = a(x)$ and $\overline{p}(x) = b(x)$ we have an R-probability.

Remark 76 (A P-Box defines an R-probability field) Let $A = (-\infty, x]$ and let two functions $\underline{F}(A) \leq \overline{F}(A)$ be given such that

$$PB = \{F \mid \underline{F}(A) \leq F(A) \leq \overline{F}(A), \ A \text{ measurable}, \ F(A) = Cum \Pr(A)\}.$$

If we let

$$a^- (A) = \underline{F}(A)$$
$$a^+ (A) = \overline{F}(A),$$

with the assumption that there exists a distribution in PB [11], then a P-Box is an R-probability field (IVPM). If $\underline{F}(A)$ and $\overline{F}(A)$ are (cumulative) distributions, then PB is a F-probability field.

The significance of these observations is that interval-valued probabilities are quite general. In addition, belief and plausibility distributions over nested sets are IVPMs. And given a random set, we can generate a belief and plausibility pair and conversely.

Remark 77 Belief and plausibility distributions are also interval-valued probability measures. To see this, let $A \subseteq X, A \in \sigma_X$. Since

$$Bel(A) + Bel(A^C) \leq 1 \leq Pl(A) + Pl(A^C). \qquad (2.20)$$

and any probability Pr has the property that

$$Pr(A) + Pr(A^C) = 1$$

then

$$Bel(A) + Bel(A^C) \leq Pr(A) + Pr(A^C) = 1 \leq Pl(A) + Pl(A^C).$$

On the other hand, we have,

$$Bel(A^C) \leq Pr(A^C) \leq Pl(A^C),$$
$$Bel(A) \leq Pr(A) \leq Pl(A),$$

so that (2.20) is satisfied. Moreover, there exists such a probability given Bel(A), for any $A \in \sigma_X$. If we define $Pr(A) \equiv Bel(A)$, then we have

$$Bel(A) = Pr(A) \leq Pl(A).$$

Likewise, if we define $Pr(A) \equiv Pl(A)$, then

$$Bel(A) \leq Pr(A) = Pl(A).$$

Remark 78 Given a random set, a belief/plausibility pair can be constructed.

$$Bel(A) = \sum_{B \in \mathcal{F} | B \subseteq A} m(B)$$

and

$$Pl(A) = \sum_{B \in \mathcal{F} | B \cap A \neq \emptyset} m(B).$$

Moreover, if $\mathcal{F} = \{A_1, A_2, ..., A_K\}$, a finite collection of subsets in the family $\mathcal{F}$ such that they are nested, $A_1 \subseteq A_2 \subseteq ... \subseteq A_K$, with $A_0 = \emptyset$ and $A_{K+1} = X$, we have

$$\emptyset = A_0 \subseteq A_1 \subseteq A_2 \subseteq ... \subseteq 7A_K \subseteq A_{K+1} = X,$$

then the belief and plausibility measures generated by the random set is a possibility and necessity measure respectively. The set of all probabilities generated by a random set $(\mathcal{F}, m)$, for a finite family $\mathcal{F}$, has the form

$$M_{RS} = \left\{ Pr \mid Pr(A) = \sum_{k=1}^{K} m(A_k) \, Pr^{\,k}(A), \, A \in \sigma_X \right\}$$

where
$$\text{Pr}^{\,k} \in \{\text{Pr}(A) = 1, A_k \subseteq A\}.$$

The representation M_{RS} of a random set can also generate another belief and plausibility measure as follows [18].

$$Bel(A) = \inf_{\text{Pr} \in M_{RS}} \text{Pr}(A),$$

$$Pl(A) = \sup_{\text{Pr} \in M_{RS}} \text{Pr}(A).$$

Moreover, if $\mathcal{F}$ is a family of singletons, then $A_k = \{x\}$ so that

$$Pl(\{x\}) = \sum_{B \in \mathcal{F} | B \cap \{x\} \neq \varnothing} m(B) = m(\{x\}).$$

That is,

$$Pl(x) = m(x).$$

Likewise,

$$Bel(\{x\}) = \sum_{B \in \mathcal{F} | B \subseteq \{x\}} m(B) = m(\{x\}),$$

and

$$Bel(x) = m(x).$$

Therefore in the case of singleton families,

$$Bel(x) = Pl(x) = \text{Pr}(x).$$

Let a measure $\mu : \sigma_X \to [0, 1]$ be an onto function where $X = \{x_1, ..., x_K\}$. Thus,

$$\mu(x_k) = \alpha_k, 0 < \alpha_k \leq 1, \forall x_k \in X,$$

and $\exists\, x_i, x_j$ such that $\mu(x_i) = 0$ and $\mu(x_j) = 1$. Without loss of generality assume that
$$0 = \alpha_1 \leq \alpha_2 \leq ... \leq \alpha_K = 1$$

by renumbering if necessary. It may happen that $\exists\, k, l, 1 \leq k \leq k + l \leq K$ such that

$$\alpha_k = \alpha_{k+1} = ... = \alpha_{k+l}.$$

In this case, use another notation for those α's. For example, suppose we have K_1 such sets of α's. Define

$$0 = \beta_1 < \beta_2 < ... < \beta_{K_1} = 1$$

where

$$\beta_1 = \alpha_1 = ... = \alpha_{l_1}$$
$$\beta_2 = \alpha_{l_1+1} = ... = \alpha_{l_2}$$
$$\vdots$$
$$\beta_{K_1} = \alpha_{l_{k_1}+1} = ... = \alpha_K.$$

Let

$$F(\beta_k) = \left\{ x_{l_{k-1}+1}, x_{l_{k-1}+2}, \ldots, x_{l_{k-1}+K} \right\}, k = 1, ..., K_1$$

be the β_k-level set. Then μ is equivalent to the random set $(F \in \mathcal{F}, m)$ where m is given by

$$m(F(\alpha_k)) = \beta_k - \beta_{k-1}, k = 1, ..., K_1, \beta_0 = 0.$$

Theorem 79 *A random set generates a belief, plausibility measure and a belief, plausibility measure can be translated into a random set.*

Proof It is clear from Remark 78 that a random set $(F \in \mathcal{F}, m)$ generates a belief, plausibility measure. Moreover, it was proved by Shafer [20] that

$$m(A) = \sum_{B|B \subseteq A} (-1)^{|A-B|} Bel(B).$$

This means that a belief, plausibility can be translated into a random set.   ∎

Remark 80 A P-Box can be translated into a random set (see [12]). In special cases, probability intervals can be translated into random sets (see [28] and Chap. 3).

2.3.1   Possibility Interval, Generalized Uncertainty Arithmetic

Possibility intervals are essentially represented by upper and lower bounding functions. The development of the associated arithmetic can be done alpha-level by alpha-level at each domain value x as in the usual way of doing interval arithmetic. However, when one represents functions as being constraint interval functions, the arithmetic is perform on the upper and lower functions in function space rather than in interval space. These definitions, discussions, and development are left to Chap. 4 where they are presented in the context of where they are used.

2.4 Summary

This section defined ten types of uncertainty entities. These ten provide a rich toolkit for the analysis of uncertainty data that arises from models in which there is an incomplete set of information. Our interest is to use these for constraint sets and objective functions of optimization whose models require generalized uncertainty analysis. As we shall see, each of these ten (under some assumptions) can be transformed to IVPMs. Subsequent chapters will consider generalized uncertainties either as pairs of possibility and necessities or as IVPMs. From these pairs in the body or rim of an optimization problem, we will develop methods of optimization.

2.5 Exercises

Exercises 81 Prove that given a possibility measure defined by (2.6)–(2.8) and the necessity measure defined by (2.9) that

$$Nec(\emptyset) = 0, \, Nec(X) = 1,$$
$$Nec(A \cap B) = \min\{Nec(A), Nec(B)\}.$$

Exercises 82 Prove Proposition 42.

Exercises 83 Prove that a trapezoidal fuzzy interval is a possibility distribution.

Exercises 84 Given random set data, show how to get an equivalent IVPM.

Exercises 85 Given a probability interval data set, show how to get an equivalent IVPM.

Exercises 86 Given a belief/plausibility data set, show how to get an equivalent IVPM data set.

Exercises 87 An entity X has realizations x_1, x_2, x_3, x_4 where $pr(x_1) \in \left[\frac{1}{3}, \frac{5}{6}\right]$, $pr(x_2) \in \left[\frac{1}{3}, \frac{1}{2}\right]$, $pr(x_3) \in \left[\frac{1}{6}, \frac{5}{6}\right]$ and $pr(x_4) \in \left[\frac{1}{4}, \frac{2}{3}\right]$. This type of information is a random set information or not? Verify your answer.

References

1. B. Bede, *The Mathematics of Fuzzy Sets and Fuzzy Logic* (Springer, Berlin, 2012)
2. G.J. Klir, B. Yuan, *Fuzzy Sets and Fuzzy Logic: Theory and Applications* (Prentice Hall, New Jersey, 1995)
3. D. Dubois, H. Prade, Gradualness, uncertainty and bipolarity: making sense of fuzzy sets. Fuzzy Sets Syst. **192**, 3–24 (2012)

4. K.D. Jamison, W.A. Lodwick, Interval-Valued Probability Measures, *UCD/CCM Report No. ???* (2004)
5. R.E. Moore, *Methods and Applications of Interval Analysis* (SIAM, Philadelphia, 1979)
6. R.E. Moore, R.B. Kearfott, M.J. Cloud, *Introduction to Interval Analysis* (Society for Industrial and Applied Mathematics, Philadelphia, 2009)
7. D. Dubois, H. Prade, Random sets and fuzzy interval analysis. Fuzzy Sets Syst. **42**(2), 1987, 87–101 (1991)
8. D. Dubois, E. Kerre, R. Mesiar, H. Prade, Chapter 10: Fuzzy interval analysis, in *Fundamentals of Fuzzy Sets*, ed. by D. Dubois, H. Prade (Kluwer Academic Press, 2000), pp. 483–581
9. D. Dubois, H. Prade, *Possibility Theory* (Plenum Press, New York, 1988)
10. K. Weichselberger, The theory of interval-probability as a unifying concept for uncertainty. Int. J. Approx. Reason. **24**, 149–170 (2000)
11. S. Ferson, V. Kreinovich, R. Ginzburg, K. Sentz, D.S. Myers, in *Constructing Probability Boxes and Dempster-Shafer Structures*. Sandia National Laboratories, Technical Report SAND2002-4015, Albuquerque, New Mexico (2003)
12. S. Destercke, D. Dubois, E. Chojnacki, Unifying practical uncertainty representations: I. generalized p-boxes. Int. J. Approx. Reason. **49**, 649–663 (2008)
13. A. Neumaier, Structure of clouds (2005) (downloadable http://www.mat.univie.ac.at/~neum/papers.html)
14. S. Destercke, D. Dubois, E. Chojnacki, Unifying practical uncertainty representations: II. clouds. Int. J. Approx. Reason. **49**, 664–677 (2008)
15. A.N. Kolmogorov, Confidence limits for an unknown distribution function. Ann. Math. Stat. **12**, 461–463 (1941)
16. A.P. Dempster, Upper and lower probability induced by a multivalued mapping. Ann. Math. Stat. **38**, 325–339 (1967)
17. L.M.D. Campos, J.F. Huete, S. Moral, Probability intervals: a tool for uncertain reasoning. Int. J. Uncertain. Fuzziness Knowl.-Based Syst. **2**(2), 167–196 (1994)
18. H. Nguyen, *An Introduction to Random Sets* (Chapman & Hall/CRC, Boca Raton, 2006)
19. L.A. Zadeh, Fuzzy sets as a basis for a theory of possibility. Fuzzy Sets Syst. **1**, 3–28 (1978)
20. G. Shafer, *A Mathematical Theory of Evidence* (Princeton University Press, Princeton, 1976)
21. K.D. Jamison, W.A. Lodwick, The construction of consistent possibility and necessity measures. Fuzzy Sets Syst. **132**, 1–10 (2002)
22. W.A. Lodwick, K.D. Jamison, Interval-valued probability in the analysis of problems that contain a mixture of fuzzy, possibilistic and interval uncertainty, in *2006 Conference of the North American Fuzzy Information Processing Society, June 3-6, 2006, Montréal, Canada*, ed. by K. Demirli, A. Akgunduz. paper 327137 (2006)
23. W.A. Lodwick, K.D. Jamison, "Interval-valued probability in the analysis of problems containing a mixture of possibility, probabilistic, and interval uncertainty. Fuzzy Sets Syst. **159**(1), 2845–2858 (2008). Accessed 1 Nov 2008
24. W.A. Lodwick, R. Jafelice Motta, Constraint interval function analysis – theory and application to generalized expectation in optimization, in *Proceedings, NAFIPS 2018/CBSF V, Fortaleza, Ceará, Brazil* (2018). Accessed 4–6 July 2018
25. W.A. Lodwick, Constrained Interval Arithmetic. *CCM Report* **138** (1999)
26. D. Dubois, S. Moral, H. Prade, Semantics for possibility theory based on likelihoods. J. Math. Anal. Appl. **205**, 359–380 (1997)
27. G. Shafer, Belief functions and possibility measures, in Chapter 3 of *Analysis of Fuzzy Information, Volume 1, Mathematics and Logic*, ed. by J.C. Bezdek (CRC Press, Inc., 1987), pp. 51–84
28. P. Boodgumarn, P. Thipwiwatpotjana, W.A. Lodwick, When a probability interval is a random set. Sci. Asia **39**, 319–326 (2013)
29. P. Thipwiwatpotjana, W.A. Lodwick, A relationship between probability interval and random sets and its application to linear optimization with uncertainties. Fuzzy Sets Syst. **231**, 45–57 (2013)

Chapter 3
The Construction of Flexible and Generalized Uncertainty Optimization Input Data

3.1 Introduction

We consider that the input data for flexible optimization are given by fuzzy intervals and/or transitional set belonging associated with relations. The input data for generalized uncertainty optimization are given by one of our ten uncertainty types discussed in the previous chapter. In particular, we have the following.

1. Flexible optimization input data:

   (a) Fuzzy intervals;
   (b) Set belonging arising from relationships.

2. Generalized uncertainty input data:

   (a) Possibility intervals;
   (b) Interval-valued probabilities;
   (c) Kolmogorov, Smirnov bounds.

If we consider the above classification and that we show that set belonging result in fuzzy intervals, the above results in four types of data types:

1. Fuzzy intervals;
2. Possibility pairs;
3. Interval-valued probabilities;
4. Kolmogorov, Smirnov bounds.

This means that when we develop constructions for flexible and generalized uncertainty data inputs, we can focus on constructions that yield one of these four types with Kolmogorov, Smirnov bounds, as we mentioned, enclosing with confidence limits.

© Springer Nature Switzerland AG 2021
W. A. Lodwick and L. L. Salles-Neto, *Flexible and Generalized Uncertainty Optimization*, Studies in Computational Intelligence 696,
https://doi.org/10.1007/978-3-030-61180-4_3

3.2 Construction Flexibility/Fuzzy Distributions

Distributions for flexible constraints are fuzzy membership functions arising directly from the given fuzzy interval membership function or constructed from soft relationships. Chapter 1 presented the definition of a fuzzy interval as a membership function of a fuzzy set which is more general than a fuzzy number. In the context of transitional set belonging, a fuzzy interval semantically represents gradual membership in a set. This section focuses on gradual set belonging as found/used in flexible optimization. There are two places where data that are part of flexible optimization enter into an optimization problem:

1. In the right-hand side coefficients;
2. In the relations.

Remark 88 A fuzzy interval in the body coefficients and/or objective function coefficients are considered as coming from uncertainty in the model and not gradual set belonging. Thus, the data corresponding to these models are considered as part of generalized uncertainty data. The first fuzzy optimization models considered the objective function as a goal. When this occurs, the objective function is converted into flexible constraint and so considered either as a right-hand side coefficient or as a gradual relationship (soft relationship). Therefore, flexibility in the objective function is not considered separately from our two flexible types listed above. That is, we have two types of flexible optimization data that we will construct. Nevertheless, we will show how to construct a soft constraint from objective function goals.

3.2.1 Construction of Data for Coefficients in Right-Hand Side

The only challenge to developing a fuzzy interval representing gradual set belonging is if it is not an a-priori given membership function. We will assume that any coefficient data that is a fuzzy number or interval is given a-priori as a membership function. For example, we might be given a fuzzy number $\tilde{2}$ representing a coefficient that is "around 2". Triangular or trapezoidal fuzzy intervals are typical membership function types. That is, for the numeric values of coefficients that are modeled as fuzzy numbers/intervals, their associated function is assumed to be a-priori known and specified. How we handle this data in an flexible optimization is covered in Chap. 5.

3.2.2 Construction of Data for Flexible/Soft Relationships

There are two places where fuzzy relationships occur in an optimization model structure.

1. **Objective Function**: The first relationship is associated with what we mean by operator "optimize" in the objective function usually stated as an equality relation

$$opt \ z = f(c, x).$$

2. **Constraint:** The second relationship occurs in the description or formulation of the constraint set typically given as the equations and/or inequalities that make up the delineation of the constraint set. Logical relationships of course may be a part of the optimization problem (such as aggregation operators of *and*, *or*, *true/false*). However we will assume that these have been transformed into systems of equalities and/or inequalities.

These two types, therefore, become one type, which is discussed next. An objective function whose relationship *"optimize"* represents a fuzzy goal is modeled in terms of flexibility, that is, a fuzzy set whose construction we are interested in obtaining. For example, a fuzzy goal in the objective function might be something like, "Come as close as possible to the target value T". Suppose we are maximizing, then "Come as close as possible to the target value T" means, in the context of flexible optimization,

$$z = f(x) \tilde{\geq} T$$

where $z = f(x)$ is the objective function and $\tilde{\geq}$ is a flexible relationship. Thus, we need to give mathematical meaning to flexible relationships. To do this, the objective is transformed into a flexible constraint which we discuss subsequently in terms of how to transform a flexible constraint relationship into a fuzzy set membership function. We note that when we have an optimization problem with fuzzy goals as objective functions and it is considered as a constraint, it is often called in the literature *symmetric* since fuzzy constraints can be considered as objective function fuzzy goals and objective fuzzy goals can be considered as fuzzy constraints. Fuzzy constraints are also called *soft* constraints as we have mentioned in Chap. 1. These constraints can all be considered as gradual set belonging $x \tilde{\in} X$. The overall objective becomes maximizing set belonging for each constraint.

Consider a linear programming problem

$$opt \ z = c^T x$$

$$\text{subject to: } \sum_{j=1}^{n} a_{ij} x_j \leq, =, \geq \ b_i, 1 \leq i \leq m,$$

and let the objective function be given by the zeroth row of a matrix representation of dimension $(m + 1) \times n$ *when* we have a flexible objective. The constraint matrix

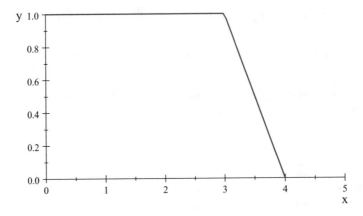

Fig. 3.1 Flexible constraint $A_{io}x \leq 3$ where $d_i = 1$

for a flexible optimization problem where $\in$ is gradual becomes A_{ij}, where $0 \leq i \leq m$, $1 \leq j \leq n$, is given by

$$A_{io}x \tilde{\leq} b_i, i \in \{0, \ldots, m\}$$

$$\updownarrow$$

$$\max \alpha$$

$$A_{io}x \leq b_i + (1 - \alpha)d_i, i \in \{0, \ldots, m\}$$

$$0 \leq \alpha \leq 1. \tag{3.1}$$

Here, the d_i is an a-priori given flexibility bound and is part of the known input data of the problem, that is, it is part of the data set. In the way we have presented it above, which is the original approach of [4], the fuzzy membership function representing this flexibility is *linear*. For example, suppose that we have

$$y_i = A_{io}x \leq 3, x \geq 0$$

and $d_i = 1$, then for a linear flexibility, the membership function for this flexible relationship is depicted as in Fig. 3.1. Here, d_i is given data and the translation between set belonging, $\mu(x) = 1$, and not belonging to the set, $\mu(x) = 0$, must be chosen. Depicted is a *linear* transition between belonging and non-belonging. What does this mean? Suppose we are constrained by a total of three units but we have the ability to add one more unit to the constraint (for example, borrowing extra worker hours from another department to take care of a sudden onset of customers) but our preference is to be limited by 3 units and, at the same time, we are absolutely limited by 4 units (for example, by law).

Example 89 (*Flexible Optimization*) Suppose we have the real-valued linear programming problem

$$\max z = x_1 + 2x_2$$
$$x_1 + 4x_2 \leq 16$$
$$x_1 + x_2 \leq 10$$
$$x_1, x_2 \geq 0.$$

Now, suppose the flexibility in the first constraint is 3 and in the second constraint is 1. According to our notation, $d_1 = 3$ and $d_2 = 1$. Moreover, suppose our goal is to come as close as possible to $\hat{z} = 15$ where the optimal value of the given problem is $z^* = 12$, $x_1^* = 8$, $x_2^* = 2$. This means that $d_0 = 3$ and the objective function becomes (exceed as much as possible but never go below the optimal value of 12)

$$x_1 + 2x_2 \geq 15 - (1 - \alpha)3.$$

Thus, our model translates into

$$\max \hat{z} = \alpha$$
$$-x_1 - 2x_2 \leq -15 + (1 - \alpha)3$$
$$x_1 + 4x_2 \leq 16 + (1 - \alpha)3$$
$$x_1 + x_2 \leq 10 + (1 - \alpha)1$$
$$0 \leq \alpha \leq 1.$$

Simplifying,

$$\max \hat{z} = \alpha + 0x_1 + 0x_2$$
$$3\alpha - x_1 - 2x_2 \leq -12$$
$$3\alpha + x_1 + 4x_2 \leq 19 \qquad (3.2)$$
$$\alpha + x_1 + x_2 \leq 11$$
$$x_1, x_2 \geq 0 \text{ and } 0 \leq \alpha \leq 1.$$

The maximum is at $\alpha = \frac{5}{14} = 0.357...$, $x_1 = \frac{115}{14} = 8.2...$, $x_2 = \frac{17}{7} = 2.4...$ with an objective function value of $z = \frac{183}{14} = 13.071...$ which is more than the non-flexible optimal value of $z^* = 12$ as expected. We have violated our original first constraint by 1.9... and our second constraint by 0.6... which is what an α different than 1 gives.

The process to transform general fuzzy/soft constraint into a flexible optimization problem with linear membership functions is the following where we use *max* for *opt* to be concrete.

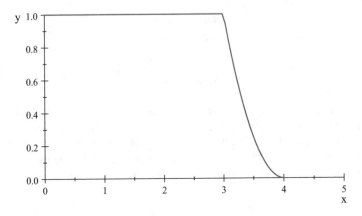

Fig. 3.2 Convex membership function $d_i = 1$

$$\max z = f(c, x) \tilde{\geq} T$$
$$\text{subject to:} g_i(a, x) \tilde{\leq} b_i, \, 1 \leq i \leq m$$

$$\Updownarrow$$

$$\max z = \alpha$$
$$f(c, x) \geq T - (1 - \alpha)d_0$$
$$g_i(a, x) \leq b_i + (1 - \alpha)d_i, \, 1 \leq i \leq m,$$
$$0 \leq \alpha \leq 1, \tag{3.3}$$

where the d_i, $i = 0, \ldots, m$, are the a-priori given flexibilities. Recall that we are just focusing on flexibility of the relationships. The rim coefficients c and body coefficients could also be fuzzy intervals but we present this subsequently. As mentioned, the use of linear membership functions was the original way translations of a fuzzy relationship into a fuzzy interval membership function in the context of flexibility were made (see [4]). However, any continuous non-increasing function $h(\alpha)$ could be used where

$$h(\alpha), \, 0 \leq \alpha \leq 1, \text{ where } h(1) = 0, h(0) = 1$$

and (for a flexible less than or equal constraint) we would have

$$g_i(x) \leq b_i + h_i(\alpha)d_i. \tag{3.4}$$

For example, a convex $h(\alpha) = (1 - \alpha)^2$ would result in

$$g_i(x) \leq b_i + (1 - \alpha)^2 d_i$$

and is depicted in Fig. 3.2, with $A_{i\circ}x \leq 3$ where $d_i = 1$.

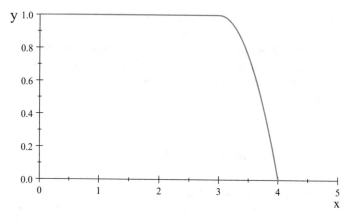

Fig. 3.3 Concave membership $d_i = 1$

A concave $h(\alpha) = -\alpha^2 + 1$ with $d_i = 1$ is depicted in Fig. 3.3 with $A_{i\circ}x \le 3$ where $d_i = 1$.

Remark 90 Note that when we use a nonlinear membership function our linear programming problem becomes a nonlinear programming problem. This is why a linear membership function is usually used.

Remark 91 One can perhaps see that a flexible optimization model like (3.3) with a large number of constraints, in general, is not Pareto optimal. This has been mentioned in Chap. 1 and will be discussed further in Chap. 5.

Definition 92 A feasible point, $\bar{x}$, is said to be **Pareto optimal,** for a maximization optimization problem, if there does not exist another feasible point, x, such that $f(x) \ge f(\bar{x})$.

This section presented the construction of data that is used in flexible optimization. There are fuzzy intervals arising in the right-hand sides and in relationships/soft constraints. Soft constraints, in turn, are modeled as fuzzy intervals. Linear fuzzy intervals are typical since a flexible linear programming remains a linear program whereas a non-linear fuzzy interval results in a nonlinear program.

3.3 Construction of Generalized Uncertainty Distributions

Generalized uncertainty distributions arise from information deficiency as modeled by one of the four types (fuzzy interval, possibility pairs, IVPMs, Kolmogorov/Smirnov bounds (KSBs)). Pairs of distributions that guarantee to bound the underlying unknown distribution will be called *enclosures*. Since the semantic of the data in this section is tied to *lack of information* our construction for this section generates *pairs of distributions*. Given the construction of a pair of distributions, the use of generalized uncertainty data in optimization can be one of three types, (1) The pair yielding optimistic and pessimistic optimization, (2) Any distribution within the pair yielding an approximate optimization, or (3) Both simultaneously yielding a minimax regret or penalized optimization. Let us begin looking at a construction that we have seen before.

3.3.1   Intervals and Fuzzy Intervals

Fuzzy intervals tied to information deficiency have a construction that we have looked at previously. Here we review the construction via a simple example again. We emphasize that we are constructing pairs of distributions.

Example 93 Suppose all we know is that the support of an unknown probability function lies in the interval $[1, 3]$. This is partial information and that the enclosure for the cumulative distribution functions are:

$$pos_2(x) = \begin{cases} 0 \text{ if } -\infty < x < 1 \\ 1 \text{ if } \quad 1 \leq x \leq \infty \end{cases}$$

and

$$pos_1(x) = \begin{cases} 0 \text{ if } -\infty < x < 3 \\ 1 \text{ if } \quad 3 \leq x \leq \infty \end{cases}.$$

Clearly, these two extreme distributions arise from cumulative probabilities whose function is the Dirac Delta Function at the endpoints of the interval. That is, with just the support being all the information we have, and with no other information, the probability could all be at $x = 1$ or at $x = 3$. Upper and lower functions are in fact a pair of possibility distribution functions and themselves cumulative probability functions, that enclose all possible cumulative probabilities whose probability distribution function has support $[1, 3]$. One type of optimization under generalized uncertainty works with this type of enclosure, a pair in this case which are also cumulative distributions. Another type of optimization under generalized uncertainty uses single distribution function

$$int_{[1,3]} = \begin{cases} 1 \text{ for } x \in [1, 3] \\ 0 \text{ for } x \in (-\infty, 1) \cup (3, \infty) \end{cases}.$$

What is the difference? Which of these two types would we use for a given optimization problem? This is discussed in Sect. 3.3.2.1. However, the reader is already alerted to the challenges that uncertainty data presents and therefore in applying generalized uncertainty to optimization, the user needs to be attentive to the requisites and objective of the analysis.

3.3.2 Possibility Pairs Distributions

There are at least four broad approaches one can use to construct possibility and necessity distributions as was mentioned in Chap. 2. They are:

1. Through a fuzzy set membership function [5],
2. Through *fuzzy intervals* or, in general, *normalized fuzzy sets* over decomposable domains [6],
3. Axiomatically from *fuzzy measures* g that satisfy

$$g(A \cup B) = \max\{g(A), g(B)\}, \tag{3.5}$$

References [7, 8],
4. Consistent probability based possibility [9].

The first approach is via a fuzzy membership function which was the way possibility was originally proposed by Zadeh [5] and which we introduced in Chap. 1. The second is via fuzzy intervals and is almost the same as the first since a fuzzy interval is a particular membership function. In addition to just accepting the fuzzy membership function either as a Zadeh possibility or as a given fuzzy interval as a possibility, fuzzy interval membership function as a possibility can be used to generate a upper possibility and a lower possibility pair of bounding functions as in Fig. 3.5. This pair has an additional property that they are themselves cumulative distributions so they enclose all cumulative probabilities between this pair that have the corresponding support as that of the fuzzy interval that generated these bounding cumulative distributions. This is different from the original way possibility was introduced by Zadeh. What needs to be kept in mind are three things before a fuzzy membership can be a valid possibility.

1. The domain of the membership function must be decomposable (see [6]).
2. The semantics of possibility are tied to information deficiency, lack of information, partial information and *not* to gradual set belonging.
3. Possibility always models existent entities, that is, for our case of real-valued domains. That is, our possibility distributions will always need to have at least one value whose possibility is one.

As long as our domain is a subset of the real number line it is decomposable. Moreover, any construction from the Zadeh approach via fuzzy sets must have at least one membership value of 1 for at least one domain value. That is, the fuzzy membership from which the possibility is constructed must be normalized. Thus, if we derive a possibility distribution from a fuzzy set and the membership function is normalized and the domain of the membership function is the set of real numbers, the only thing that needs to be verified is that the semantics of the entity modeled by the membership function is tied to information deficiency (lack of information, partial information). Let us next look at the difference between a single possibility versus a possibility pair (or a possibility/necessity pair).

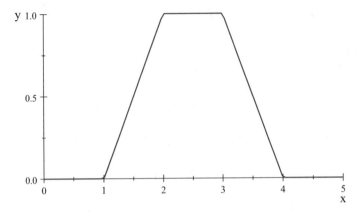

Fig. 3.4 Single possibilistic distribution

3.3.2.1   Single Possibility Versus Possibility/Necessity Pairs Used in Optimization

Single possibility distributions may be used to generate pairs of bounding distributions. Let us try to clarify what is at times confusing when we try to use possibility in mathematical analysis, optimization in our case. Let us consider the following. Suppose we have the ith constraint right-hand side value b_i lying between 1 and 4. That is, we definitely know that this constraint cannot be less than 1 (for example, we are required to have one unit present, say one hour of time) and we do not physically have enough resources to obtain more than 4 units. However, we know that it is most desirable to have between 2 and 3 units available at all times where the desirability between 1 and 2 (because of loss of revenue say) is linear and the undesirability (because of extra expense for overtime, say) is linear between 3 and 4 units. This is depicted in Fig. 3.4. This single possibilistic distribution can be used directly in an optimization problem as will be shown in Chap. 4. However, if the data represents partial information like the following example, then we do not have a single distribution but a pair.

Example 94 Suppose that to satisfy customers we must have at least 1 h of a clerk available. To satisfy our human power constraint, we can have no more than 4 h. Customers and management are completely satisfied with between 2 and 3 h of service with customers satisfaction increasing linearly for availability of between 1 and 2 h while management has a decreasing linear satisfaction between 3 and 4 h. We will in fact, deliver between 1 and 4 h but it will depend on other constraints and what our objective function will be (from a customer point of view or a management point of view or somewhere between). From the point of view of management, we would try to deliver the dashed-red (left) possibility distribution $1/2/\infty/\infty$ of Fig. 3.5. From the customers' point of view, we would try to deliver the dashed-blue (right) possibility distribution $3/4/\infty/\infty$ of Fig. 3.5 whereas the incomplete information is represented

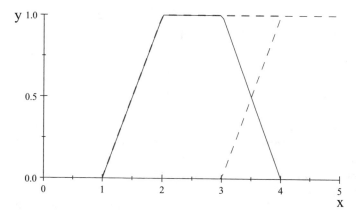

Fig. 3.5 Possibilistic/necessity distribution pair

by the trapezoid 1/2/3/4. An optimization problem could use the trapezoid, one of the two bounding possibility distributions depending of the point of view, a single distribution between using a recourse-like methods, or both together using a penalty (or a minimax regret).

Remark 95 The method that we develop will construct an upper and lower possibility distributions. A probability-based possibility construction depends on our ability to construct underlying set of nested subsets that we call a possibility nest (defined below). Interval-valued probability construction we develop will obtain upper and lower bounding (cumulative distribution) functions over Borel sets, subsets of a σ-algebra on the real line, which are more general than possibility nests. If we are able to construct a possibility nest from our partial information, it is a much simpler way to construct a bounding pair of distributions. The power of our development of interval-valued probabilities is that from possibility/necessity pairs constructed from possibility nests, we can extend the distributions to Borel sets maintaining the enclosure. From a practical point of view, we construct distributions from nests.

3.3.2.2 Possibility Distribution Construction Via Fuzzy Membership Function

The way possibility was first developed by Zadeh [5] was from a fuzzy membership function as has been mentioned. In this case, given a membership function $\mu(x)$, we immediately have the possibility $pos(x) = \mu(x)$ *provided that* $x \in \mathbb{R}$ (the domain is decomposable), it is normalized, and the semantic is that of lack of information. Recall our example of meeting a tall female at the airport where the evaluation of females arriving at the baggage claim was done via the possibility of their being *tall* using the fuzzy membership function of *tall*, and the membership function we use is normalized, that is, there exists at least one female in existence, x^*, which is certainly tall. We can use this as a single distribution or construct a pair depending on the purpose to which the data will be put. Recall that a fuzzy number is a fuzzy membership function.

3.3.2.3   Possibility Via Fuzzy Interval Function Construction

A most useful approach to constructing possibility distributions to be used in generalized uncertainty optimization is from fuzzy intervals assuming their semantics are associated with information deficiency. This is because when numeric coefficients are formed from lack of information, they are generally modeled as fuzzy intervals. The construction of possibility pairs from fuzzy intervals can be visualized in Fig. 3.5. Formally we have the following construction. We note that a possibility distribution function is generically denoted $pos(x)$ and we add further identifiers when we have different possibility function distributions that need to be distinguished.

Remark 96 Given a fuzzy interval whose membership function is $\mu(x)$ whether given by the Zadeh approach, an a-priori given fuzzy interval, or by other methods. Let

$$C = \{x \,|\, \mu(x) = 1\}$$
$$L = \{x \,|\, \mu(x) \text{ is non-decreasing}\}$$
$$U = \{x \,|\, \mu(x) \text{ is non-increasing}\}$$
$$L^* = L - C$$
$$U^* = U - C$$
$$\underline{x} = \sup L^*, \bar{x} = \inf U^*$$

where the interval $[a, b] \subset \mathbb{R}$ is the support of $\mu(x)$, that is, for $x \le a, x \ge b$, $\mu(x) = 0$, and $\mu(x) > 0$, for $a < x < b$. We also have that $a \le \underline{x} \le x^* \le \bar{x} \le b, x^* \in C$. Since there exists x^* such that $\mu(x^*) = 1$, $\underline{x}$ and $\bar{x}$ are well-defined. Now define the upper possibility and lower necessity as follows:

$$pos_2(x) = \begin{cases} 0 \text{ for } x \le a \\ \mu(x) \text{ for } a < x \le \underline{x} \ , \\ 1 \text{ for } x > \underline{x} \end{cases} \tag{3.6}$$

and

$$pos_1(x) = \begin{cases} 0 \text{ for } x \le \bar{x} \\ 1 - \mu(x) \text{ for } \bar{x} < x \le b \ . \\ 1 \text{ for } x > b. \end{cases} \tag{3.7}$$

It is clear that $pos_1(x)$ and $pos_2(x)$ are cumulative probability distributions. Moreover, any non-decreasing function $pos(x)$ such that

$$pos_1(x) \leq pos(x) \leq pos_2(x), \tag{3.8}$$

is also a possibility and a cumulative distribution. Moreover, it is also clear that they satisfy the axioms of a possibility measure. We leave this as an exercise. Note that pos_1 is really the dual necessity of the fuzzy interval possibility μ as we have shown in Chap. 2.

3.3.2.4 Axiomatic Construction

The construction from axioms of possibility theory found in [7] or [8] is often a theoretical tool rather than a practical way to construct possibility distributions and not directly relevant to our construction approach since to use the axioms, one needs to already have a possibility distribution g and this section is about how to construct the distribution g. The main use to which we put the axioms is to verify that indeed we have a possibility measure. For example, the axioms were used to prove, in Remark 96, that the upper function (3.6) and lower function (3.7) were possibility measures which means we do in fact have a pair of possibility distributions.

3.3.2.5 Probability Based Possibility Construction

This section is derived from [9]. The consistent possibility and necessity measures we construct are derived from nested families of measurable sets. In addition, the consistent possibility and necessity measures on the range space of a measurable function can be constructed from consistent possibility and necessity measures on the domain using the extension principle. This is particularly useful since the objective function, and constraint set definitions described with function relationships $g(x) \leq b$, for example, in optimization problems whose input domain (parameters for example) is uncertain and mapped to a range space (via the objective function) onto a complete lattice (the real numbers for example). Thus, if we have a consistent possibility/necessity pair generated by the uncertainty in the domain (via an unknown probability distribution for example), it induces a consistent possibility/necessity pair in the range space. It is with these induced consistent possibility/necessity pairs in the range space that can be used in optimization.

Remark 97 On the real number line, it is clear that we can pick a countable set of nested sets $[a, b)$ that will generate a σ-algebra on any closed interval. Since all closed intervals can be transformed into $[0, 1]$, let our σ-algebra be generated by $\{[0, 1 - \frac{1}{n}), n = 1, 2, \ldots\} \cup [0, 1]$ where, of course, $[0, 0) = \emptyset$.

We first list basic definitions including the definition of a consistent possibility or necessity measure. Then some basic theorems concerning consistent possibility and necessity measures are given. These include:

1. A method for constructing consistent possibility measures;
2. A proof that the set of consistent possibility measures completely determines the measure;
3. A proof that the extension of consistent possibility measures is consistent;
4. A proof that the set of such extended possibility measures completely determines the induced measure.

What (1)–(4) mean from a practical point of view is that if we have (incomplete) probability data and we are able to construct a possibility nest of subsets in the domain $\mathbb{R}$ rather than on elements of the domain, then there is a verifiable way to construct possibility and necessity distributions that enclose the induced measure in the range space. Note that we are assuming that the probability is *not* known on the *elements* of a set, as, for example, in the data given in Example 94. If we had probability data on the elements of the domain, we would have the complete probability density function as a single-valued function and would not need the construction below.

Definition 98 Given a measure space $(X \subset \mathbb{R}, \sigma_X, \mu)$ and $\mathcal{P}(X)$ the power set of X, then a set function $Pos : \mathcal{P}(X) \to \mathbb{R}_\infty$ is called a **possibility measure consistent with the measure** μ and the set function $Nec : \mathcal{P}(X) \to R_\infty$ is called a **necessity measure consistent with the measure** μ if
(1) $\forall E \in \sigma_X$, a Borel sets on X

$$Nec\,(E) \leq \mu(E) \leq Pos(E),$$

(2) $\forall \{A_\alpha\}_{\alpha \in \Lambda} \subseteq \mathcal{P}(X)$,

$$Pos\,(\cup_{\alpha \in \Lambda} A_\alpha) = \sup\{Pos\,(A_\alpha) \mid \alpha \in \Lambda\},$$

and

$$Nec\,(\cap_{\alpha \in \Lambda} A_\alpha) = \inf\{Nec\,(A_\alpha) \mid \alpha \in \Lambda\}.$$

(3) Lastly,

$$Pos\,(X) = \mu\,(X) = Nec\,(X)$$

and

$$Pos\,(\emptyset) = \mu(\emptyset) = Nec\,(\emptyset) = 0.$$

Remark 99 Parts (2) and (3) of the definition insure that we have a possibility/necessity measure whereas (1) is what we mean by consistency.

The level sets of possibility distributions are used quite frequently. Since for our case, $\mu\,(X) = 1$, the level sets are called α−cuts [8].

There are two other approaches to constructing probability-based possibility. The first approach is associated with data where a set of probabilities are given and we do not know which of the set is the correct one. The second approach is associated with a construction knowing probability on sets. The interpretation we adopt for possibility in our setting of optimization is that they can bound on a set of cumulative probability distributions. Thus, given a pair of possibility measures, $Pos_1 \leq Pos_2$, and a given random outcome A, we want to construct a set with these possibilities such that

$$CDFBnd = \{Cum\,Pr \mid Pos_1(A) \leq Cum\,Pr(A) \leq Pos_2(A)\}. \qquad (3.9)$$

That Eq. (3.9) can be constructed is clear since we can start with a fuzzy interval that define the distributions associated with Pos_1, Pos_2 as in (1.3) and (1.4). This set, $CDFBnd$, (3.9), is non-empty since by construction, the distributions associated with Pos_1 and Pos_2 are CDFs. This interpretation of possibility theory provides a method of combining possibility, intervals, probability, and interval-valued probability into a single mathematical framework which we will do subsequently. Thus we concentrate on construction where we start with probabilities and generate possibilistic enclosures.

3.3.2.6 Possibility from a Set of Probabilities

The easiest way to obtain probability-based possibility (3.9) associated with a fuzzy interval is the following. Let $\mu(x)$ be the membership function of a fuzzy interval and $Pr(A)$ be probability measures for which

$$\sup_{a \in A} \mu(a) \geq Pr(A). \qquad (3.10)$$

Define

$$Pos(A) = \sup_{a \in A} \mu(a). \qquad (3.11)$$

$Pos(A)$ so defined, is a possibility measure since

$$Pos(\emptyset) = \sup_{a \in \emptyset} \mu(a) = 0$$
$$Pos(X) = \sup_{a \in X} \mu(a) = 1$$

and all fuzzy intervals have a non-empty core. Moreover,

$$Pos(A \cup B) = \sup_{x \in A \cup B} \mu(x)$$
$$= \sup_{x \in A} \mu(x) \text{ or } \sup_{x \in B} \mu(x)$$
$$= \max\{\sup_{x \in A} \mu(x), \sup_{x \in B} \mu(x)\}$$
$$= \max\{Pos(A), Pos(B)\}.$$

$Pos(A)$ so defined satisfies (2.6) and (2.7) and thus a possibility.

Given this definition of possibility (3.11), the following holds.

$$Pos(A) = \sup_{a \in A} \mu(a) \geq \Pr(A) \text{ for all measurable } A.$$

$$Nec(A) = 1 - Pos(A^C) \text{ or}$$

$$Pos(A^C) = 1 - Nec(A).$$

Now,

$$Pos(A^C) \geq \Pr(A^C) = 1 - \Pr(A) \text{ or}$$

$$\Pr(A) \geq 1 - Pos(A^C) = Nec(A). \text{ Thus,}$$

$$Nec(A) \leq \Pr(A) \leq Pos(A). \tag{3.12}$$

That is, when we define a possibility by (3.11), we generate a possibility/necessity dual pair of bounding measures for all probabilities that satisfy (3.10). Note that we have a possibility consistent with probability.

3.3.2.7   Probability-Based Possibility Via Possibility Nests

Definition 100 ([9]) Let $p : S \rightarrow [0, 1]$ be a regular possibility distribution function with associated possibility measure Pos and necessity measure Nec. Then p is said to be **consistent** with respect to the random variable X if for all measurable sets A,

$$Nec(A) \leq \Pr(X \in A) \leq Pos(A)$$

where

$$Pos(A) = \sup_{a \in A}\{p(a)\} \geq \Pr(A)$$

and Nec is defined to be

$$Nec(A) = 1 - Pos(A^C).$$

This results in

$$Pos(A^C) \geq \Pr(A^C)$$
$$= 1 - \Pr(A) \text{ or}$$
$$\Pr(A) \geq 1 - Pos(A^C)$$
$$= Nec(A),$$

which means that we have a possibility/necessity pair consistent with the random variable X.

Definition 101 Let p and n be possibility and necessity distribution functions respectively. Define the α-**cuts** of p and n to be the sets

$$p^\alpha = \{x \mid p(x) \geq \alpha\} \text{ and } n^\alpha = \{x \mid n(x) \leq \alpha\}$$

and the **strong** α-**cuts** to be the sets

$$p^{\alpha+} = \{x \mid p(x) > \alpha\} \text{ and } n^{\alpha+} = \{x \mid n(x) < \alpha\}.$$

The following definition will be used in the method for constructing consistent possibility and necessity measures as follows.

Definition 102 Let (X, σ_X, μ) be a measure space. A collection of measurable sets,

$$PN = \{E_r \mid r \in S \subseteq R_\infty\}, \, S \text{ an ordered indexing set,}$$

will be called a **possibility nest** if it satisfies the following properties:
(1) $r < s \Rightarrow E_r \subset E_s$ (i.e. PN is nested)
(2) $X, \emptyset \in PN$
(3) $\forall t \, \exists E_r \in PN$ such that

$$\mu(E_r) = \mu\left(\cup_{\mu(E_s) < t} E_s\right)$$

and $\forall t \, \exists E_u \in PN$ such that

$$\mu(E_u) = \mu\left(\cap_{\mu(E_s) > t} E_s\right).$$

Remark 103 Property (3) insures that there are no "gaps" in PN in the sense that the set $T = \{\mu(E_r) \mid r \in S\}$ must be closed. This follows since for any limit point t of T we can construct a sequence of points $\{t_n\} \subseteq T$ converging to t either from above or below. Then property (3) insures there is an $E_r \in PN$ such that $\mu(E_r) = t$.

Example 104 Consider $X = [0, 1]$ with the Lebesgue measure and

$$PN = \left\{\emptyset, X, \left\{\left[0.25 + \frac{1}{n}, 0.75 - \frac{1}{n}\right] \mid n = 4, \ldots, \infty\right\}\right\}.$$

Then

$$\mu\left(\cup_{\mu(E_s) < 0.5} E_s\right) = \mu\left(\cup_{n=4}^{\infty}\left[0.25 + \frac{1}{n}, 0.75 - \frac{1}{n}\right]\right) = 0.5$$

but $\nexists E_r \in PN$ such that $\mu(E_r) = 0.5$ so that PN fails property (3) and is not a possibility nest.

Example 105 For $X = [0, 1]$ with Lebesgue measure the collection of sets

$$PN = \{X, (1 - \alpha, 1] \mid \alpha \in [0, 1]\}$$

satisfies properties 1, 2 and 3 and is a possibility nest where $(1, 1] = \emptyset$.

That the conversion of possibility to a necessity measure preserves consistency on a finite measure space is given next. Recall, from Remark 97, that a possibility nest exists for closed intervals of $\mathbb{R}$.

Theorem 106 *Given a possibility measure Pos consistent with a finite (probability) measure μ, then the set function Nec,*

$$Nec(A) = \mu(X) - Pos(A^c),$$

is a necessity measure consistent with μ. If Nec is a consistent necessity measure then the set function,

$$Pos(A) = \mu(X) - Nec\left(A^c\right),$$

is a possibility measure consistent with μ.

Proof

$$\forall E \in \sigma_X$$

(1) $Nec(E) = \mu(X) - Pos(E^c) \leq \mu(X) - \mu(E^c) = \mu(E).$

(2) $Nec\left(\bigcap_{k \in \Lambda} A_k\right) = \mu(X) - Pos\left(X - \bigcap_{k \in \Lambda} A_k\right)$

$$= \mu(X) - Pos\left(\bigcup_{k \in \Lambda} (X - A_k)\right)$$

$$= \mu(X) - \sup_{k \in \Lambda} Pos(X - A_k)$$

$$= \inf_{k \in \Lambda} (\mu(X) - Pos(X - A_k))$$

$$= \inf_{k \in \Lambda} \left(\mu(X) - Pos(A_k^c)\right)$$

$$= \inf_{k \in \Lambda} (Nec(A_k)).$$

(3) $Nec(X) = \mu(X) - Pos(\emptyset) = \mu(X)$ and $Nec(\emptyset) = \mu(X) - Pos(X) = 0$

The proof for the second part is completely analogous.   ■

Remark 107 Since a probability measure has the property that $\mu(X) = 1$, given a consistent possibility, the usual definition of the dual necessity,

$$Nec(A) = 1 - Pos(A^c)$$

is consistent by Theorem 106. To our knowledge, the first to define necessity as the dual to possibility were Dubois and Prade [7]. They observed that, from the axioms and Zadeh's initial approach to possibility [5], if $Pos(A)$ were known, $Pos(A^c)$ is not a known value unlike what occurs in probability. What one needs is a dual necessity measure and in this case $Pos(A^c) = 1 - Nec(A)$. Thus, a more complete characterization of the uncertainty than what a possibility measure models is one that has its associated dual measure, its necessity measure. Recall that in probability theory, we have, for $A \cap B = \emptyset$, $\Pr(A \cup B) = \Pr(A) + \Pr(B)$. In possibility theory we have $Pos(A \cup B) = \max\{Pos(A), Pos(B)\}$ regardless of whether or not the intersection is empty. Moreover, in probability theory $\Pr(A \cup A^c) = 1 = \Pr(A) + \Pr(A^c)$ so that

$$\Pr(A^c) = 1 - \Pr(A).$$

In possibility theory, the equivalent would be,

$$Pos(A^c) = 1 - Nec(A).$$

The general method for construction of consistent possibility and necessity measures is developed from a possibility nests. In essence consistent possibility and necessity measures are, respectively, special outer and inner measures on this possibility nest.

Theorem 108 (see [9]) *Let* (X, σ_X, μ) *be a measure space and*

$$PN = \{E_r \mid r \in S \subseteq \mathbb{R}_\infty\}$$

be a possibility nest. Then the set functions $Pos, Nec : P(X) \to R_\infty$ *defined by*

$$Pos(A) = \inf\{\mu(E_r) \mid A \subseteq E_r, E_r \in PN\}$$

and

$$Nec(A) = \sup\{\mu(E_r) \mid E_r \subseteq A, E_r \in PN\}$$

are possibility and necessity measures consistent with μ.

This theorem establishes a method to construction a possibility (necessity) measure and from this to construct the possibility (necessity) distribution. We must, however, have a possibility nest on which the underlying probability measure is defined.

We next turn our attention to functions over domains on which we have constructed consistent possibility and necessity distributions and construct consistent measures induced on the range space of a measurable function which will be useful in optimization. The result of the next theorem establishes the fact that the extension of consistent possibility and necessity measures is itself consistent.

Theorem 109 (see [9]) *Let* (X, σ_X, μ) *be a measure space and* (Y, σ_Y) *a measurable space. Let* $f : X \to Y$ *be an* σ_X-*measurable function and let* ν *be the measure on* Y *defined by*

$$\nu(E) = \mu\left(f^{-1}(E)\right) \forall E \in \sigma_Y.$$

Let p_X *and* n_X *be possibility and necessity distribution functions for possibility and necessity measures* Pos_X *and* Nec_X *consistent with* μ. *Then the functions*

$$p_Y, n_Y : Y \to \mathbb{R}_\infty$$

defined by

$$p_Y(y) = \sup\{p_X(x) \mid f(x) = y\}$$

(where we define $\sup \emptyset = 0$*) and*

$$n_Y(y) = \inf\{n_X(x) \mid f(x) = y\}$$

(where we define $\inf \emptyset = \mu(X)$*) are possibility and necessity distribution functions for a possibility measure* Pos_Y *and necessity measure* Nec_Y *consistent with* ν*.*

The significance of this theorem is that by determining $Pos_Y(A)$ we're actually determining the infimum over the $\mu(E_r)$'s such that $f^{-1}(A) \subseteq E_r$. Similarly $Nec_Y(A)$ gives the supremum over the E_r's such that $E_r \subseteq f^{-1}(A)$.

Example 110 Suppose we are given the following,

$$PN_{poss} = \begin{cases} E_0 = \emptyset, \\ E_i = (-\infty, x_i), \ x_i = 1 + i, \ i = 1, 2, 3, 4, \\ E_5 = (-\infty, 6], \\ E_6 = \mathbb{R} \end{cases},$$

that is,

$$PN_{poss} = \{\emptyset, (-\infty, 2), (-\infty, 3), (-\infty, 4), (-\infty, 5), (-\infty, 6], \mathbb{R}\}.$$

This possibility nest enables us to obtain a consistent possibility from the data as follows.

Intervals X_i		$Pr(x \in X_i)$	Cumulative for $X_i = Pos(X_1 \cup \cdots \cup X_i)$
X_1	$(-\infty, 2)$	0.0	0.0
X_2	$[2, 3)$	0.1	0.1
X_3	$[3, 4)$	0.2	0.3
X_4	$[4, 5)$	0.3	0.6
X_5	$[5, 6)$	0.4	1.0
X_6	$[6, \infty)$	0.0	1.0

Starting at the other end, the possibility nest that leads to a necessity measure is:

$$PN_{nec} = \begin{cases} W_0 = \emptyset, \\ W_1 = (6, \infty) \\ W_i = [6 - i, \infty), \ i = 1, 2, 3, 4 \\ W_6 = \mathbb{R} \end{cases}$$

which results in the following cumulative probabilities,

Intervals W_i		Cumulative for W_i
W_1	$(6, \infty)$	0.0
W_2	$[5, \infty) = [5, 6) \cup [6, \infty)$	0.4
W_3	$[4, \infty) = [4, 5) \cup [5, \infty)$	0.7
W_4	$[3, \infty) = [3, 4) \cup [4, \infty)$	0.9
W_5	$[2, \infty) = [2, 3) \cup [3, \infty)$	1.0
W_6	$(-\infty, \infty)$	1.0

Note that for the possibility of say, $W_3 = [4, \infty) = [4, 5) \cup [5, \infty)$, we have from the above,

$$0.7 = Pos(W_3)$$
$$= Pos([4, 5) \cup [5, \infty))$$
$$= \max\{Pos([4, 5)), Pos([5, \infty))$$
$$= \max\{Pos[4, 5), 0.4\}$$

This implies that $Pos([4, 5)) = 0.7$ so that we have the following.

Z_i		$Pos(Z_i)$
Z_0	$(-\infty, \infty)$	$Pos((-\infty, \infty)) = Pos(W_6) = 1.0$
Z_1	$[2, 3)$	$Pos([2, \infty)) = Pos(W_5) = 1.0$
Z_2	$[3, 4)$	$Pos([3, \infty)) = Pos(W_4) = 0.9$
Z_3	$[4, 5)$	$Pos([4, \infty)) = Pos(W_3) = 0.7$
Z_4	$[5, 6)$	$Pos([5, \infty)) = Pos(W_2) = 0.4$
Z_5	$[6, \infty)$	$Pos([6, \infty)) = Pos(W_1) = 0.0$

Let $Z_6 = \emptyset$, so that

Intervals Y_i	$Y_i^c = Z_i \Leftrightarrow$ Pos interval	Cumulative for $Nec(Y_i) = 1 - Pos(Y_i^c)$ $= 1 - Pos(Z_i)$
$Y_0 = \emptyset$	$(-\infty, \infty) = Z_0$	$1.0 - 1.0 = 0.0$
$Y_1 = (-\infty, 2)$	$[2, \infty) \Leftrightarrow [2, \infty) = Z_1$	$1.0 - 1.0 = 0.0$
$Y_2 = (-\infty, 3)$	$[3, \infty) \Leftrightarrow [2, 3) = Z_2$	$1.0 - 0.9 = 0.1$
$Y_3 = (-\infty, 4)$	$[4, \infty) \Leftrightarrow [3, 4) = Z_3$	$1.0 - 0.7 = 0.3$
$Y_4 = (-\infty, 5)$	$[5, \infty) \Leftrightarrow [4, 5) = Z_4$	$1.0 - 0.4 = 0.6$
$Y_5 = (-\infty, 6)$	$[6, \infty) \Leftrightarrow [5, 6) = Z_5$	$1.0 - 0.0 = 1.0$
$Y_6 = (-\infty, \infty)$	$\emptyset = Z_6$	$1.0 - 0.0 = 1.0$

The resulting possibility pairs consistent with the underlying (unknown) probability is depicted by Fig. 3.6. Note that we can also consider the enclosing distributions from the point of view of interval-valued probability as well. Two cases for which probability based possibilities are easily constructed are given next.

Case 111 (*Finite non-singleton partition of the real line with no overlaps*) Let a partition of the real line be given as

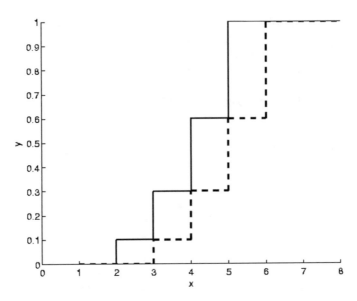

Fig. 3.6 Consistent Pos non-dual Nec wrt an unknown prob

$$P_N = \{-\infty < x_1 < \cdots < x_{N-1} < x_N < \infty\}$$
$$PN_N = \{A_1, A_2, \ldots, A_N\}$$

where

$$A_1 = (-\infty, x_1)], A_2 = [(x_1, x_2)], \ldots, A_N = [(x_{N-1}, x_N)], A_{N+1} = [(x_N, \infty)$$

and ")]", "[("mean that if) is on the right endpoint of the preceding interval, [is on the left of the next interval and if] is on the right endpoint of the preceding interval, (is on the left next interval. Moreover, the probability on each interval of the partition needs to be given. That is, we are given

$$\Pr(A_k) = p_k \geq 0, \sum_{k=1}^{N+1} p_k = 1.$$

Then,

$$pos(x) = \sum_{k=1}^{n} p_k, x \in A_n,$$
$$nec(x) = \sum_{k=1}^{n-1} p_k, x \in A_n.$$

This is the construction that was used to obtain the two distributions of Fig. 3.6.

Case 112 (*Probability based possibility of finite non-singleton partition of the real line with overlaps*) Suppose we have

$$A_1 = (-\infty, x_1)], \ A_2 = [(x_1, x_2)], \ldots, A_N = [(x_{N-1}, x_N)], \ A_{N+1} = [(x_N, \infty)$$

with corresponding probabilities, where x_j is not necessarily less than or equal to x_{j+1}, that is, $A_j \cap A_{j+1}$ may be non-empty with more than one element,

$$\Pr(A_k) = p_k \geq 0, \ \sum_{k=1}^{N+1} p_k = 1.$$

However, assume that for at least one pair of subintervals, $x_k > x_{k+1}$. In this case, construct a new interval $[(x_{k+1}, x_k)]$, renumbering and adjusting the open/closed endpoints in such a way that there is no overlap. The question, then is what is the probability associated with the new interval? We now have three subintervals (without renumbering at this point)—$(x_{k-1}, x_{k+1}]$, $(x_{k+1}, x_k]$, and $(x_k, x_{k+2}]$ where we have chosen the open/closed sequence to simplify the notation. The probabilities we have are $\Pr((x_{k-1}, x_k]) = p_k$ and $\Pr((x_k, x_{k+1}]) = p_{k+1}$. With no information to assist us in the assignment of probabilities, one way to approximate the probabilities is to use the interval length ratios to assign probabilities. That is, let

$$d_k = x_k - x_{k-1},$$
$$d_{k+1} = x_k - x_{k+1},$$
$$d_{k+2} = x_{k+2} - x_{k+1}$$

and

$$f_1 = \frac{d_{k+1}}{d_k},$$
$$f_2 = \frac{d_{k+1}}{d_{k+2}}$$

so that

$$\Pr((x_{k-1}, x_{k+1}]) = (1 - f_1) \, p_k,$$
$$\Pr((x_{k+1}, x_k]) = f_1 p_k + f_2 p_{k+1},$$
$$\Pr((x_k, x_{k+2}]) = (1 - f_2) \, p_{k+1}.$$

Then, one proceeds in the same way as in the non-overlapping partition case.

Example 113 Suppose we have slight alteration of the data given in Example 110 as follows:

Intervals X_i		Pr$(x \in X_i)$	Cumulative for $x \in X_i$
X_1	$(-\infty, 2)$	0.0	0.0
X_2	$[2, 3.5)$	0.1	0.1
X_3	$[3, 4)$	0.2	0.3
X_4	$[4, 5)$	0.3	0.6
X_5	$[5, 6]$	0.4	1.0
X_6	$(6, \infty)$	0.0	1.0

This results in the non-overlapping data set

Intervals X_i		Pr$(x \in X_i)$	Cumulative for $x \in X_i$
X_1	$(-\infty, 2)$	0.0	0.0
X_2	$[2, 3)$	0.066...	0.066...
X_3	$[3, 3.5)$	0.1	0.166...
X_4	$[3.5, 4)$	0.1333...	0.3
X_5	$[4, 5)$	0.3	0.6
X_6	$[5, 6]$	0.4	1.0
X_7	$(6, \infty)$	0	1.0

and the upper possibility and lower necessity are constructed as before.

Remark 114 To obtain guaranteed enclosure then each overlapping interval is considered as a possibility and necessity pair with respect to the interval where the endpoints of the intervals are Dirac-Delta probability distributions as we did with respect to $\mu_{[a,b]}(x)$ in constructing $pos_{[a,b]}(x)$ and $nec_{[a,b]}(x)$.

3.3.3 Interval-Valued Probabilities: Construction of Interval-Valued Probability Measures from Upper/Lower Cumulative Probabilities

Typically, one is given probability bounds with the data. We next show how interval probabilities are constructed on the Borel sets from lower and upper bounding cumulative distribution functions defined on a set of intervals, for example a possibility nest. That is, probability data is given on *sets* (intervals in this case), not on elements (singletons). What we seek to do in this section is to develop in a formal manner a structure given two cumulative distributions $F^u(x) \geq F^l(x)$ defined on a subset of intervals, show how they can be extended to enclosing cumulative distributions of an unknown random variable $X \in \mathcal{M}(X^u, X^l)$ where

$$\mathcal{M}(X^u, X^l) = \{X \mid \forall x\ F^u(x) \geq \Pr(X \leq x) \geq F^l(x)\},$$

and for any Borel set $A \subset \mathbb{R}$, $\Pr(A) = \Pr(X \in A)$, which is our unknown probability density function whose (unknown) cumulative distribution we are enclosing, either approximately via Kolmogorov–Smirnov bounds (KSBs) or completely via probability-based possibility (PBP), for example. Above, Examples 110 and 113 informally introduced probabilities on intervals that generated pairs of enclosing distributions. Here, we will begin by assuming that we know the bounding cumulatives over basic simple sets (possibility nests for example). Specifically, our cumulative distributions are consistent bounds for members of the family of sets

$$\mathcal{I} = \{(a, b], (-\infty, a], (a, \infty), (-\infty, \infty), \emptyset \mid a < b\}.$$

For $I = (-\infty, b]$ it is clear by definition that

$$\Pr(I) \in \left[F^l(b), F^u(b)\right] = i_m(I)$$

where i_m denotes our interval-valued probability function. For $I = (a, \infty)$, let

$$\Pr(I) \in \left[1 - F^u(a), 1 - F^l(a)\right].$$

For $I = (a, b]$, since $I = (-\infty, b] - (-\infty, a]$, and considering minimum and maximum probabilities in each set, let

$$\Pr(I) \in \left[\max\left\{F^l(b) - F^u(a), 0\right\}, F^u(b) - F^l(a)\right].$$

Therefore, if we extend the definition of F^u, and F^l by defining

$$F^u(-\infty) = F^l(-\infty) = 0,$$

and

$$F^u(\infty) = F^l(\infty) = 1,$$

we can make the following general definition.

Definition 115 For any $I \in \mathcal{I}$, if $I \neq \emptyset$, define

$$i_m(I) = \left[a^-(I), a^+(I)\right] = \left[\max\left\{F^l(b) - F^u(a), 0\right\}, F^u(b) - F^l(a)\right]$$

where a and b are the left and right endpoints of I. Otherwise set

$$i_m(\emptyset) = [0, 0].$$

Remark 116 Note that with this definition

$$i_m((-\infty, \infty)) = \left[\max\left\{F^l(\infty) - F^u(-\infty), 0\right\}, F^u(\infty) - F^l(-\infty)\right]$$
$$= [1, 1],$$

which matches our intuition and thus, it is easy to see that $\Pr(I) \in i_m(I) \ \forall I \in \mathcal{I}$.

We can extend this to include finite unions of elements of $\mathcal{I}$. For example, if

$$E = I_1 \cup I_2 = (a, b] \cup (c, d]$$

with $b < c$, then we consider the probabilities,

$$\Pr((a, b]) + \Pr((c, d]),$$

and

$$1 - (\Pr((-\infty, a]) + \Pr((b, c]) + \Pr((d, \infty))),$$

(the probability of the sets that make up E versus one minus the probability of the intervals that make up the complement) by obtaining the minimum and maximum probabilities for each case as a function of the minimum and maximum of each set. The *minimum* for the first sum is

$$\max\left(0, F^l(d) - F^u(c)\right) + \max\left(0, F^l(b) - F^u(a)\right),$$

and the *maximum* is

$$F^u(d) - F^l(c) + F^u(b) - F^l(a).$$

The minimum for the second is

$$1 - \left(F^u(\infty) - F^l(d) + F^u(c) - F^l(b) + F^u(a) - F^l(-\infty)\right)$$
$$= F^l(d) - F^u(c) + F^l(b) - F^u(a)$$

and the maximum is

$$1 - \left(\max\left\{0, F^l(\infty) - F^u(d)\right\} + \max\left\{0, F^l(c) - F^u(b)\right\}() + \max\left\{0, F^l(a) - F^u(-\infty)\right\}\right)$$
$$= F^u(d) - \max\left\{0, F^l(c) - F^u(b)\right\} - F^l(a).$$

This gives, for $E = I_1 \cup I_2$,

$$\Pr(E) \geq \max \begin{cases} F^l(d) - F^u(c) + F^l(b) - F^u(a) \\ \max\left\{0, F^l(d) - F^u(c)\right\} + \max\left\{0, F^l(b) - F^u(a)\right\} \end{cases},$$

and

$$\Pr(E) \leq \min \begin{cases} F^u(d) - \max\left\{0, F^l(c) - F^u(b)\right\} - F^l(a) \\ F^u(d) - F^l(c) + F^u(b) - F^l(a) \end{cases},$$

so

$$\Pr(E) \in [\max\{0, F^l(d) - F^u(c)\} + \max\{0, F^l(b) - F^u(a)\},$$
$$F^u(d) - \max\{0, F^l(c) - F^u(b)\} - F^l(a)].$$

The final line is arrived at by noting that

$$\forall x, y \; F^l(x) - F^u(y) \leq \max\{0, F^l(x) - F^u(y)\}.$$

Remark 117 Note that there are two extreme cases. For $E = (a, b] \cup (c, d]$. For $F^u(x) = F^l(x) = F(x) \; \forall x$, then, as expected,

$$\Pr(E) = F(d) - F(c) + F(b) - F(a) = \Pr((a, b]) + \Pr((c, d]),$$

that is, it is the probability measure. For $F^l(x) = 0 \; \forall x$,

$$\Pr(E) \in \left[0, F^u(d)\right],$$

that is, it is a possibility measure for the possibility distribution function $F^u(x)$.

Next, let

$$\mathcal{E} = \left\{\cup_{k=1}^{K} I_k \mid I_k \in \mathcal{I}\right\}.$$

That is, $\mathcal{E}$ is the algebra of sets generated by I. It is clear that every element of E has a unique representation as a union of the minimum number of elements of $\mathcal{I}$. Stated differently, every element of $\mathcal{E}$ is a union of disconnected elements of $\mathcal{I}$. Note also that $\mathbb{R} \in \mathcal{E}$ and $\mathcal{E}$ is closed under complements.

Assume $E = \cup_{k=1}^{K} I_k$ and $E^c = \cup_{j=1}^{J} M_j$ are the unique representations of E and E^c in $\mathcal{E}$ in terms of elements of $\mathcal{I}$. Then, considering minimum and maximum possible probabilities of each interval, it is clear that

$$\Pr(E) \in [\max\left(\Sigma_{k=1}^{K} a^-(I_k), 1 - \Sigma_{j=1}^{J} a^+(M_j)\right), \min\left(\Sigma_{k=1}^{K} a^+(I_k), 1 - \Sigma_{j=1}^{J} a^-(M_j)\right)].$$

This can be made more concise using the following result.

Theorem 118 (see [10]) *If $E = \cup_{k=1}^{K} I_k$ and $E^c = \cup_{j=1}^{J} M_j$ are the unique representations of E and $E^c \in \mathcal{E}$, then $\Sigma_{k=1}^{K} a^-(I_k) \geq 1 - \Sigma_{j=1}^{J} a^+(M_j)$, and $\Sigma_{k=1}^{K} a^+(I_k) \geq 1 - \Sigma_{j=1}^{J} a^-(M_j)$.*

This means that i_m can be extended to $\mathcal{E}$ which is stated in the next theorem.

Theorem 119 (see [10]) *For any $E \in \mathcal{E}$, let $E = \cup_{k=1}^{K} I_k$, and $E^c = \cup_{j=1}^{J} M_j$ be the unique representations of E and E^c in terms of elements of $\mathcal{I}$, respectively. If*

$$i_m(E) = \left[\Sigma_{k=1}^{K} a^-(I_k), 1 - \Sigma_{j=1}^{J} a^-(M_j)\right],$$

then $i_m : \mathcal{E} \to Int_{[0,1]}$ is an extension of $\mathcal{I}$ to $\mathcal{E}$ and is well-defined. In addition,

$$i_m(E) = \left[\inf\left\{\Pr(X) \in E \mid X \in \mathcal{M}\left(X^u, X^l\right)\right\}, \sup\left\{\Pr(X) \in E \mid X \in \mathcal{M}\left(X^u, X^l\right)\right\}\right].$$

The family of sets, $\mathcal{E}$, is a ring of sets generating the Borel sets $\mathcal{B}$. For an arbitrary Borel set S, then it is clear that

$$\Pr(S) \in \left[\sup\left\{a^-(E) \mid E \subseteq S, E \in \mathcal{E}\right\}, \inf\left\{a^+(F) \mid S \subseteq F, F \in \mathcal{E}\right\}\right]$$

This leads us to the following theorem.

Theorem 120 (see [10]) *Let* $i_m :.\mathcal{B} \to [0, 1]$ *be defined by*

$$i_m(A) = \left[\sup\left\{a^-(E) \mid E \subseteq A, E \in \mathcal{E}\right\}, \inf\left\{a^+(F) \mid A \subseteq F, F \in \mathcal{E}\right\}\right]$$

Then i_m *is an extension from* $\mathcal{E}$ *to* $\mathcal{B}$, *and it is well-defined.*

Putting these theorems together, we have the following.

Theorem 121 (see [10]) *The function* $i_m : \mathcal{B} \to Int_{[o,1]}$ *defines an interval-valued probability field on the Borel sets and*

$$i_m(B) = \left[\inf\left\{\Pr(X \in B) \mid X \in \mathcal{M}\left(X^u, X^l\right)\right\}, \sup\left\{\Pr(X \in B) \mid X \in \mathcal{M}\left(X^u, X^l\right)\right\}\right],$$

that is, $\mathcal{M}\left(X^u, X^l\right)$ *defines a structure.*

Remark 122 In summary, given probabilities on subsets of a Borel set, an IVPM can be constructed.

3.3.4   Construction from Kolmogorov–Smirnov Statistics

Constructing bounds on an unknown distribution via Kolmogorov–Smirnov statistics (KSS) is accomplished beginning with being given incomplete statistical data to construct a single-valued distribution. Secondly, we must be given an a-priori confidence limit required or chosen for the problem. Then several theorems are used to obtain the bounds on the unknown distribution whose statistics is manifest in the data, the partial statistics. We start by stating the Kolmogorov theorem.

The Kolmogorov Theorem ([11]) Given the empirical distribution function F_n for n independent distributed observations x_i where

$$F_n(x) = \frac{1}{n}\sum_{i=1}^{n} \chi_{x_i}(x)$$

and

$$\chi_{x_i}(x) = \begin{cases} 1 \text{ if } x_i \leq x \\ 0 \text{ otherwise} \end{cases}.$$

Given incomplete statistics that would allow us to obtain a single-valued distribution and a predetermined confidence limit, the Kolmogorov–Smirnov statistic states that for a given distribution F (hypothesized to be the actual underlying distribution described by the data),

$$D_n = \sup_x \| F_n(x) - F(x) \|$$

where

$$\sqrt{n} D_n \leq 1 - \alpha$$

and $1 - \alpha$ is the confidence interval and D_n was obtained by Smirnov and is now obtained from tables.

The theorem is used to test whether or not a cumulative distribution that we think will describe our data meets a confidence interval criteria given by Kolmogorov's Theorem as computed via Smirnov's tables found in any standard advanced statistical text. What we will do is the opposite. Given a data set, we want to obtain upper and lower bounds on the unknown cumulative distribution postulated to have generated our given data set so the confidence interval on the unknown distribution is what we specify. We re-state the Kolmogorov–Smirnov Theorem in the context of constructing upper and lower cumulative distributions for a given data set and an a-priori desired confidence level. In particular, we have the following theorem.

Theorem 123 (See, for example, https://onlinecourses.science.psu.edu/stat414) *A* $100(1 - \alpha)\%$ *confidence band for the unknown cumulative distribution function* $F(x)$ *generated by statistical data is given by* $F_L(x)$ *and* $F_U(x)$ *where*

$$F_L(x) = \begin{cases} 0, & \text{if } F_n(x) - d \leq 0 \\ F_n(x) - d, & \text{if } F_n(x) - d > 0 \end{cases}, \tag{3.13}$$

$$F_L(x) = \begin{cases} F_n(x) + d, & \text{if } F_n(x) - d < 1 \\ 1, & \text{if } F_n(x) + d \geq 1 \end{cases}, \tag{3.14}$$

and d is selected so that

$$P(D_n \geq d) = \alpha$$
$$D_n = F_n(x) - F(x).$$

Proof (*Also see any standard advanced mathematical statistics text*) Given that d was selected so that $P(D_n \geq d) = \alpha$, this implies, from the definition of D_n, and the probability of complementary events, that

$$P(\sup_x | F_n(x) - F(x) | \leq d) = 1 - \alpha$$

If the largest of the absolute values of $F_n(x) - F(x)$ are less than or equal to d, then all the absolute values of $F_n(x) - F(x)$ must be less than or equal to d. That is,

$$P(|F_n(x) - F(x)| \le d \ \forall x) = 1 - \alpha$$

or

$$P(-d \le F_n(x) - F(x) \le d \ \forall x) = 1 - \alpha$$
$$P(-F_n(x) - d \le -F(x) \le -F_n(x) + d \ \forall x) = 1 - \alpha$$
$$P(F_n(x) - d \le F(x) \le F_n(x) + d \ \forall x) = 1 - \alpha.$$

Note that it is possible that in the lower limit $F_n(x) - d$ we could have a negative value and the upper limit we could have a value more than 1. This is taken care of by writing the cumulative probabilities as (3.13) and (3.14).   ■

Example 124 (*See* https://onlinecourses.science.psu.edu/stat414) Each person in a random sample of $n = 10$ was asked about X, the daily time wasted at work doing non-work related activities, such as surfing the internet, facebook postings and emailing friends. The resulting data, in minutes, are as follows:

$$108 \ 112 \ 117 \ 130 \ 111$$
$$131 \ 113 \ 113 \ 105 \ 128 \ .$$

We want to use this data to construct a 95% confidence band for the unknown cumulative distribution function F(x). We start by ordering the x values. The formulas for the lower and upper confidence limits tell us that we need to know d and $F_n(x)$ for each of the 10 data points. Because the confidence level is 0.05 and the sample size n is 10, the table of Kolmogorov–Smirnov Acceptance Limits (published in standard tables) tells us that d = 0.41. Thus, we have the following:

k	ORDERED x	$F_n(x) = \frac{k}{10}$
1	105	0.1
2	108	0.2
3	111	0.3
4	112	0.4
5	113	0.5
6	113	0.6
7	117	0.7
8	128	0.8
9	130	0.9
10	131	1.0

To calculate the lower limit, we have

$$F_L(x) = F_n(x) - d$$
$$= F_n(x) - 0.41.$$

The first four data points: $0.1 - 0.41 = -0.31, 0.2 - 0.41 = -0.21, 0.3 - 0.41 = -0.11$, and $0.4 - 0.41 = -0.01$ are negative. Therefore, these are assigned lower limit of 0. Similarly, in calculating the upper limit,

$$F_U(x) = F_n(x) + d = F_n(x) + 0.41,$$

the upper limit would be greater than 1 for the last five data points: $0.6 + 0.41 = 1.01, 0.7 + 0.41 = 1.11, 0.8 + 0.41 = 1.21, 0.9 + 0.41 = 1.31$, and $1.0 + 0.41 = 1.41$. Therefore, these are assign the upper limit of 1. This results in

k	ORDERED x	$F_n(x) = \frac{k}{10}$	$F_L(x)$	$F_U(x)$
1	105	0.1	0	0.51
2	108	0.2	0	0.61
3	111	0.3	0	0.71
4	112	0.4	0	0.81
5	113	0.5	0.09	0.91
6	113	0.6	0.19	1.0
7	117	0.7	0.29	1.0
8	128	0.8	0.39	1.0
9	130	0.9	0.49	1.0
10	131	1.0	0.59	1.0

There are tables that given an α will give the d. We present a stock market example of the use of the Kolmogorov–Smirnov statistic approach.

Example 125 (*derived by Dr. K. D. Jamison*) Kolmogorov–Smirnov bounds for an example based on 86 stock data from 1926–2012 which were aggregated for a ± 0.025 confidence interval is derived as follows. The first four, fiftieth, and eighty-sixth (sorted low return to high) stock data from which the bounds were computed are given in the table below. The column Stock(k) is the aggregated return from the lowest to highest and $Fn(k)$ is simply $\frac{k}{86}$.

Stock(k)	$Fn(k) = \frac{k}{86}$
$S_1 = -0.3729$	0.01163
$S_2 = -0.371$	0.02326
$S_3 = -0.3654$	0.03488
$S_4 = -0.3625$	0.04651
$\vdots$	$\vdots$
$S_{50} = 0.1433$	0.5814
$\vdots$	$\vdots$
$S_{86} = 0.5341$	1.0000

Next, a grid for the x-axis was chosen to be between -0.400 (to the left of the lowest return) and 0.550 (to the right of the highest return) in 0.010 increments for the calculations of the upper and lower KSBs and is labeled "Return" though the graph is given in increments of 7% (and 4% for bonds). "Upper" and "Lower" are the upper and lower calculated KSBs with a ± 0.025 interval (taken from the Smirnov table of values where $n = 86$). The column "Normal" is the normal distribution for the stock data where the computed mean for the stock data was 8.8% and the resultant standard deviation was 20.4% for this stock data. That is, the normal column contains the values at the grid points (the values of -0.40 to 0.55 in increments of 0.01) of a normal distribution whose mean is 8.8% and whose standard deviation is 20.4%. The column "Empirical" is the value corresponding to our $\frac{k}{86}$ division. For example, from the normal column, we see if stock data is indeed a normal distribution whose mean is 8.8% and standard deviation is 20.4%, with probability 0.008375, we will get a return of -40%. Here at -40% we have no stocks (since the lowest return for this data set is -37.29% and we get a $Fn(k) = \frac{1}{86}$ only at -0.37 (-37%) where we have one stock less than or equal to -37% so that $Fn(k) = \frac{1}{86}$ which corresponds to the column "Empirical". The lower is an optimistic view of the returns while the upper is a pessimistic view. Thus, for example, in the 9% return row below, the lower bound says that a return of 9% or less will occur with probability of 0.307 whereas the upper says that this will occur with a probability of 0.600. Note also, that for this data, the normal is between the computed lower and upper KSB cumulative distributions.

Return	Lower	Empirical	Upper	Normal
−0.400	−	−	0.147	0.008375
−0.390	−	−	0.147	0.009561
−0.380	−	−	0.147	0.010892
−0.370	−	0.023	0.170	0.012381
⋮	⋮	⋮	⋮	⋮
0.090	0.307	0.453	0.600	0.503911
⋮	⋮	⋮	⋮	⋮
0.550	0.853	1.000	1.000	0.988235

The resulting graph is given in Fig. 3.7. In addition, K. D. Jamison computed Kolmogorov–Smirnov bounds for 86 bond returns from 1926–2012 in the same way that was done for stocks. The resulting graph is given in Fig. 3.8.

3.3.5   Belief and Plausibility Construction

Construction of possibilities from belief and plausibility rely on nested focal elements. However, if we have a data set with nested elements, we have a possibility nest in which case we would use probability based possibility which is developed next. The interested reader can consult [3, 8, 9, 12] for further information of how

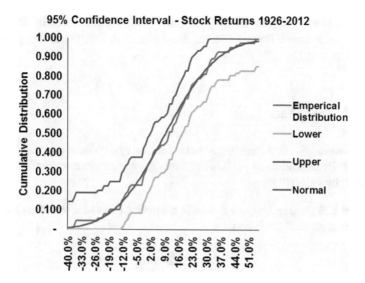

Fig. 3.7   95% confidence interval—stocks 1926–2012

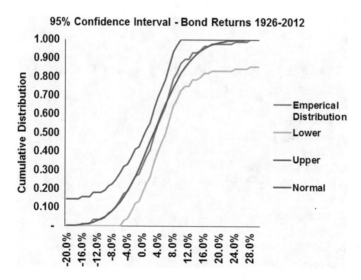

Fig. 3.8   95% confidence interval—bonds 1926–2012

to transform belief and plausibility pairs into possibility and necessity pairs. If we have belief and plausibility over nested focal elements, they are in fact possibility and necessity pair.

3.3.6   Random Sets

This section contains two algorithms that transform two types of generalized uncertainty, all P-Boxes and some probability intervals, into random sets which is shown by the next two algorithms.

Algorithm 126   A Generalized P-Box is a Special Case of a Random Set: (see [13]) Given a generalized P-Box

$$PB = [\underline{F}, \overline{F}]$$

on the nested sets

$$\emptyset = A_0 \subset A_1 \subset \cdots \subset A_n \subset A_{n+1} = X$$

and bounds

$$\alpha_i = \underline{F}(A_i) \leq \Pr(A_i) \leq \overline{F}(A_i) = \beta_i, \forall i = 0, 1, \ldots, n + 1,$$

for k = 1, ..., n+1 do
build partitions $A_i \backslash A_{i-1}$
Rank α_i, β_i increasingly
for k = 0, ..., 2n+1
rename α_i, β_i and γ_l such that
$\alpha_0 = \gamma_0 = 0 \leq \gamma_1 \leq \cdots \leq \gamma_l \leq \cdots \leq \gamma_{2n} \leq 1 = \gamma_{2n+1} = \beta_{n+1}$
end for k
Define focal set $E_0 = \emptyset$
for k = 1, ..., 2n+1
if $\gamma_{k-1} = \alpha_i$ then
$E_k = E_{k-1} \cup F_{i+1}$
if $\gamma_{k-1} = \beta_i$ then
$E_k = E_{k-1} \backslash F_i$
set m(E_k) = $\gamma_k - \gamma_{k-1}$.
end for k

3.3.7 Probability Intervals and Random Sets

We next provide another algorithm for constructing the random set. In this case we construct random sets from probability intervals. If the data is in the form of a reachable probability interval satisfying the hypotheses of Theorem 128, the algorithm can be used to obtain a random set which contains the same information as the probability interval.

Algorithm 127 Transformation of a Probability Interval to a Random Set

Let the hypotheses of Theorem 128 where $X = \{x_1, x_2, \ldots, x_n\}$, $n \geq 3$, $\Pr(\{x_i\}) \in [l_i, u_i]$, and set $m(A) = 0$ $\quad \forall A \in P(X)$, the power set of X, $sumL = 0$, $sumU = 0$, $c = 0$, and max $= 0$.

for i=1,...,n
sumL = sumL + l_i
sumU = sumU + u_i
$m(\{x_i\}) = l_i$
$i := i + 1$
end

Step 1: Check for proper probability interval
if sumL > 1 or sumU < 1
return "data not probability interval" (Stop)
else

   for i $= 1$,...,n
Step 2: Check to see if data is a reachable probability interval
if sumL $- l_i + u_i > 1$ or sumU $- u_i + l_i < 1$
return "data not reachable probability interval" (Stop)
end

Step 3: Check for the conditions of Theorem 128
if sumL $- l_i + u_i < 1$
$c := c + 1$
$ln_c = i$
end
$i := i + 1$
end
if c > 2
return "data does not satisfy the hypothesis of Theorem 128" (Stop)
elseif c $= 2$

Step 4: Construct the random set
max $= sumL - l_{ln_1} - l_{ln_2}$
if max $< 1- u_{ln_1} - u_{ln_2}$
max $= 1- u_{ln_1} - u_{ln_2}$
end
$m(X \setminus \{\{x_{ln_1}\}, \{x_{ln_2}\}\}) = $ max $-$ sumL $+ l_{ln_1} + l_{ln_2}$
$m(X \setminus \{\{x_{ln_1}\}\}) = -$ max $+ 1 - u_{ln_1} - l_{ln_2}$
$m(X \setminus \{\{x_{ln_2}\}\}) = -$ max $+ 1 - u_{ln_2} - l_{ln_1}$

$m(X) = \max - 1 + u_{\ln_1} + u_{\ln_2}$
elseif c = 1
$m(X \setminus \{\{x_{\ln_1}\}\}) = 1 - u_{\ln_1} - \text{sumL} + l_{\ln_1}$
$m(X) = u_{\ln_1} - l_{\ln_1}$
else
$m(X) = 1 - \text{sumL}$
return $m(A), \forall A \in P(X)$
end

Probability intervals are not always generalizable to random sets and vice versa. However, there is a special case of a reachable probability intervals that can be translated into a random set.

Theorem 128 ([1]) *Let $X = \{x_1, x_2, \ldots, x_n\}, n \geq 3$. A reachable probability*

$$PI = \{[l_k, u_k] \,|\, 0 \leq l_k \leq u_k \leq 1, k = 1, 2, \ldots, n\}$$

can be expressed as a random set if there are at most two indices k_1, k_2 such that

$$\sum_{j \neq k_1} l_j + u_{k_1} < 1$$

and

$$\sum_{j \neq k_2} l_j + u_{k_2} < 1.$$

3.4   Summary

This chapter developed the data input side of an optimization model associated with both flexible and generalized uncertainty optimization models. Specifically, this chapter looked at how to construct fuzzy intervals when they are not given. Moreover, the construction of upper and lower distributions generalized uncertainties coming from partial, incomplete, deficient, and/or lack of information. Once we have membership functions in the case of flexibility and upper and/or lower distributions in the case of generalized uncertain, they are inputs to optimization models as outlined in the next chapter.

3.5   Exercises

Exercise 129 Show that the construction for $pos_1(x)$ and $pos_2(x)$ of (3.6), (3.7), and (3.8) are indeed a possibility according to the definition of possibility.

Exercise 130 Use $\mu(x)$,

Fig. 3.9 Membership
function $\mu(x)$

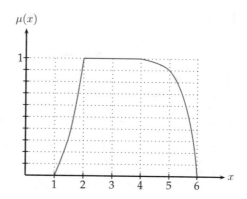

$$\mu(x) = \begin{cases} \frac{2}{3}x^2 - x + \frac{1}{3}; & 1 \leq x < 2 \\ 1; & 2 \leq x \leq 4 \\ -\frac{2}{5}x^2 + \frac{7}{2}x - \frac{33}{5}; & 4 < x \leq 6. \end{cases}$$

(see Fig. 3.9) to construct pos_1 and pos_2 as defined by (3.6), (3.7).

Exercise 131 Construct a random set from P-Box where

$$PB = \{F(x) \mid \underline{F}(x) \leq F(x) \leq \overline{F}(x)\}$$

$$\overline{F}(x) = \begin{cases} 0 \text{ for } x < 1 \\ \frac{1}{2}x - \frac{1}{2} \text{ for } 1 \leq x \leq 3 \\ 1 \text{ for } x > 3 \end{cases}$$

$$\underline{F}(x) = \begin{cases} 0 \text{ for } x < 1 \\ \frac{1}{2}x - 1 \text{ for } 2 \leq x \leq 4 \\ 1 \text{ for } x > 4 \end{cases}$$

Exercise 132 Prove that for any $A \subset B$, $\text{Bel}(A) \leq \text{Bel}(B)$. Provide an example when the equality sign happens.

Exercise 133 Given data of Exercise 124, construct the Kolmogorov, Smirnov enclosures at the 90% and 99% confidence level. You will need the Smirnov tables to do this exercise.

Exercise 134 Construct guaranteed enclosures for overlapping intervals for the data given in Exercise 113.

Exercise 135 Prove the unproved theorems contained in this chapter.

Exercise 136 If we consider an unknown entity x as the length of time the current government of a country being in power and receive a survey in the form of interval X_i as shown in the data of Example 91, answer the following questions. (a) Construct Bel/Pl cumulative distribution from the given data. (b) Do we need to partition the overlapping intervals to be able to answer (a), why or why not?

References

1. P. Boodgumarn, P. Thipwiwatpotjana, W.A. Lodwick, When a probability interval is a random set. ScienceAsia **39**, 319–326 (2013)
2. H. Nguyen, *An Introduction to Random Sets* (Chapman & Hall/CRC, Boca Raton, 2006)
3. P. Thipwiwatpotjana, W. Lodwick, Pessimistic, optimistic, and minimax regret approach for linear programs under uncertainty. Fuzzy Optim. Decis. Mak. **13**(2), 151–171 (2014)
4. H. Zimmermann, Description and optimization of fuzzy systems. Int. J. Gen. Syst. **2**, 209–215 (1976)
5. L.A. Zadeh, Fuzzy sets as a basis for a theory of possibility. Fuzzy Sets Syst. **1**, 3–28 (1978)
6. D. Dubois, The role of fuzzy sets in decision sciences: old techniques and new directions (2010)
7. D. Dubois, H. Prade, *Possibility Theory* (Plenum Press, New York, 1988)
8. G.J. Klir, B. Yuan, *Fuzzy Sets and Fuzzy Logic: Theory and Applications* (Prentice Hall, New Jersey, 1995)
9. K.D. Jamison, W.A. Lodwick, The construction of consistent possibility and necessity measures. Fuzzy Sets Syst. **132**, 1–10 (2002)
10. W.A. Lodwick, K.D. Jamison, Interval-valued probability in the analysis of problems containing a mixture of possibility, probabilistic, and interval uncertainty (2008)
11. A.N. Kolmogorov, Confidence limits for an unknown distribution function. Annu. Math. Stat. **12**, 461–463 (1941)
12. P. Thipwiwatpotjana, Linear programming problems for generalized uncertainty. Ph.D. Thesis, Department of Mathematical and Statistical Sciences, University of Colorado, 2010
13. S. Destercke, D. Dubois, E. Chojnacki, Unifying practical uncertainty representations: I. generalized p-boxes. Int. J. Approx. Reason. **49**, 649–663 (2008)

Chapter 4
An Overview of Flexible and Generalized Uncertainty Optimization

4.1 Introduction

Recall that our real-valued optimization problem is of the form,

$$opt \; z = f(c, x) : \mathbb{R}^n \times \mathbb{R}^n \to \mathbb{R}$$
$$\text{subject to:} \tag{4.1}$$
$$g(a, x) \leq b$$
$$h(d, x) = e,$$

where $g(a, x) : \mathbb{R}^n \times \mathbb{R}^n \to \mathbb{R}^{m_1}, h(d, x) : \mathbb{R}^n \times \mathbb{R}^n \to \mathbb{R}^{m_2}, b \in \mathbb{R}^{m_1}, e \in \mathbb{R}^{m_2}$, $f(c, x)$ is the *objective function*; the coefficients c are the "costs" per unit of the *variable x*, coefficients a and d are the inputs to the *body g* and h; the coefficients b and e are the *right-hand sides; and* "opt" is the *optimization operator*, that takes c, and x and maximizes, minimizes, takes a on a meaning such as "come as close as possible to …" or a meaning such as "minimize the maximum regret", and so on. To this basic optimization structure we will impose flexibility in one case and generalized uncertainty in another keeping in mind the development of these entities in the context of optimization as presented in the previous chapters. Of course, given a type of optimization model such as linear programming, the cost coefficient, body coefficients and right-hand side values have an intimate relationship via duality theory and these are often called the *rim* [4], since the cost can be transformed into the right-hand side and the right-hand side into the cost related through the transpose of the body matrix of the dual relationships. There are duality results pertaining to fuzzy and possibilistic programming, but we do not explore these. However, the interested reader can find a discussion on duality in the context of fuzzy interval programming problems in [5–7].

© Springer Nature Switzerland AG 2021
W. A. Lodwick and L. L. Salles-Neto, *Flexible and Generalized Uncertainty Optimization*, Studies in Computational Intelligence 696,
https://doi.org/10.1007/978-3-030-61180-4_4

Remark 137 To the best of our knowledge [4] coined the terms "body" and "rim" in the context of optimization and is an insightful way that structure relates to optimization analysis.

Flexible (fuzzy) and generalized uncertainty (possibilistic) optimization are two of the newest fields within optimization. The use of the designation "fuzzy optimization", in the past, meant both fuzzy (what we call flexible) and possibility (what we call generalized uncertainty) optimization. That is, in the beginning and until recently, fuzzy optimization was (and still is to some extent) used as a term to denote both fuzzy (flexible, gradual set belonging) and possibilistic (generalized uncertainty, information deficiency) adding to the confusion that fuzzy and possibilistic optimization are the same. We distinguish these here.

Fuzzy optimization began in 1970 with the publication of the seminal Bellman and Zadeh paper [8]. It took three years before the next fuzzy optimization article was published in 1973 by Tanaka, Okuda, and Asai [9, 10], with the full version coming out in 1974 [11]. These researchers seem to have been the first to realize the importance of alpha-levels in the mathematical analysis of fuzzy sets in general and fuzzy optimization in particular. The Tanaka, Okuda, Asai article was the first to operationalize the theoretical approach developed by Bellman and Zadeh. Independently, in 1974, H.-J Zimmermann presented a paper at the *ORSA/TIMS* conference in Puerto Rico [12], with the full version appearing two years later [13], that not only operationalized the Bellman and Zadeh approach, but greatly simplified and clarified fuzzy optimization, so much so that Zimmermann's approach is a standard to this day albeit, Zimmermann's approach is not, in general, Pareto optimum. In this same period, the book by Negoita and Ralescu [14] contained a description of fuzzy optimization. C. V. Negoita and M. Sularia published in 1976 a set containment approach to fuzzy optimization [15]. From these beginnings, fuzzy optimization has become a field of study in its own right with a journal devoted to the subject, *Fuzzy Optimization and Decision Making,* whose first issue came out in February of 2002.

There have been special sessions devoted solely to fuzzy/possibilistic optimization at the international fuzzy society meetings (*IFSA05*, July 2005, Beijing, China and *IFSA07* June 2007, Cancun, Mexico, *IFSA09* Lisboa, Portugal, *IFSA13* in Edmonton, Canada, *IFSA15* in Gijòn, Spain, and expected special session at *IFSA2017* in Otsu (Kyoto), Japan). *IPMU2020* in Lisboa, Portugal. In addition, there has been one special edition of the journal *Fuzzy Sets and Systems* dealing with fuzzy/possibilistic optimization [16] and an article on the beginnings of fuzzy optimization was published in 2015 [17]. At least three edited volumes on fuzzy optimization that have appeared—[18–20] and there are at least seven authored books devoted to fuzzy optimization—[7, 21–25]. While not exhaustive, this chapter will focus on the salient features of flexible and generalized uncertainty optimization. In particular, fuzzy multi-objective programming, fuzzy stochastic programming, and fuzzy dynamic programming are not discussed (see, for example [20, 23, 24, 26–29]). Moreover, since intervals (and real numbers) may be considered as fuzzy sets, interval optimization is not covered separately but considered within the family of flexible (fuzzy) or generalized uncertainty optimization depending on whether its semantics is tied to gradual set belonging or to information deficiency.

Forty-five years of fuzzy optimization research has yielded a wide-ranging set of applications which is beyond the aim of this monograph. However, the interested reader may wish to consult the bibliographies associated with [3, 20]. There have been many surveys in the area of fuzzy optimization. Among all of these, the following are noted: [3, 18, 19, 29–35].

The implementation of a flexible and generalized uncertainty optimization problem might be thought to consist of the following components, which are needed in putting together a model.

1. *DATA ACQUISITION:* Obtain the requisite inputs to the model. There must be either fuzzy relations, fuzzy goals, targets, or fuzzy intervals in order for there to be a flexible or generalized optimization problem. That is, the data of the model are the parts of its structure—$a, b, c, d, e, opt, relations$ ($\leq, =, \geq, \in$) and there must be at least one component that makes it flexible or generalized uncertainty.

   (a) Determine if there are fuzzy goals that relate to the data inputs. If there are fuzzy goals, use the fuzzy interval data and/or goals to transform these into fuzzy relationships as developed below. In this case, the whole optimization model or part is flexible optimization. This is made clear below and in Chaps. 5 and 6.
   (b) Determine if the fuzzy intervals are associated with incomplete or partial information. If so, then it is generalized uncertainty. We make this clear below and in more detail in Chap. 6.
   (c) Identify the fuzzy and possibilistic relationships ($opt, \leq, =, \in$). We discuss this in more detail below and in Chaps. 5 and 6.

2. *TRANSLATION:*

   (a) Take the fuzzy relationships and translate them into associated membership functions, fuzzy intervals. If there are deterministic equations, they remain as they are and need no transformation. This is essentially (4.24) as developed below.
   (b) For flexible/fuzzy optimization, use an aggregation operator (a t-norm) to transform the fuzzy membership function into a real-valued constraint set. This is essentially a flexible optimization delineated below by Eq. (4.33). The only choice at this point is the aggregation operator.
   (c) For generalized uncertainty optimization, take the fuzzy intervals associated with incomplete/partial information and explicitly form a functional relationship as delineated by Eq. (4.34). Also see below.
   (d) Translate the problem into a real-valued mathematical programming problem. Here there are choices to be made which we outline in the next chapters. Typically, however, for flexible optimization we would use something like Zimmermann's approach [13] whereas for generalized uncertainty, one has several approaches from "traditional" possibilistic optimization of Buckley

[30], Inuiguchi, Sakawa, Kume [36], Luhandjula [26], or Verdegay [34]. However, there are newer approaches that use robust optimization, minimax regret or penalty methods adapted to the generalized uncertainty environment that we outline in more detail subsequently.

3. *OPTIMIZE*: Once the problem is put into a real-valued optimization problem, the optimization is solved with a traditional mathematical programming method and is, in principle, straight-forward.
4. *INTERPRET*: Once an optimal is found, the interpretation of the solution is often necessary. For example, suppose one is doing a scheduling problem for an auto assembly line and the optimization model comes up with a fuzzy 23 units to be processed by the assembly line per shift. What does this mean? One does not hand the person responsible for the assembly line schedule implementation the following: *"Give me a fuzzy twenty-three units."* So, the interpretation of the solution via the appropriate semantics is part of the problem.

4.2 Flexibility and Generalized Uncertainty Translations

This section outlines some methods requisite to transform flexibility (including soft relationships) and generalized uncertainties of interest into possibility and/or necessity measures or distributions.

Let us look at two situations. (1) Suppose one's goal is to maximize the outcome of some fuzzy set, the range y for $y = \mu_A(x)$ and x in some fuzzy set A. In this case we have a fuzzy (flexible) optimization. (2) On the other hand, suppose one's goal is to maximize the valuation, the value of the distribution of possible outcomes of our choice of action where each choice generates a (possibility) distribution. It is like choosing particular stock option(s) in which to invest based on the temporal distributions of their returns of, for example, the last 25 years. In this case, one is choosing a distribution which optimizes some valuation of the returns. This is possibilistic (generalized uncertainty) optimization. We introduce these topics and their details will follow in subsequent chapters.

Relationships: The interpretation of functional and set relationships can be made transitional (fuzzy), *soft relationships*. We distinguish two types of relationships pertinent to optimization: (1) constraint relationships (equality, inequality, subsethood, belonging), and (2) operational relationship *optimize* (maximize, minimize). The first we call *soft constraint relationships* and the second *soft objective relationships*.

4.2.1 Input Coefficients

This section has taken, in part, from [37, 38]. Interval functions can be thought of arising from real-valued functions. Real functions $f(x)$ can be considered as being composed of "unknown" variable(s) x and "known" input coefficient(s) a. Thus a function may be expressed as $f(a, x)$. In the context of fuzzy intervals and generalized uncertainty, we assume that the function either has the domain as intervals (where both a and x belong to intervals) whose range is a real set, which are referred to as **interval functions**, or the function can map real values to intervals at each x, in which case are referred to as interval-valued functions. Formally we have the following definitions where we do not distinguish, at this point, between input variables and given known input coefficients.

Definition 138 An **interval function** is

$$f : x \in [x] \to f([x]), [x] \in [\underline{x}, \overline{x}] \subset \mathbb{IR}, f([x]) \subset R_{ange} \in \mathbb{R}^n$$

Definition 139 An **interval-valued function** $[F(x)]$ is given by

$$[F(x)] = [\underline{f}(x), \overline{f}(x)], x \in \mathbb{R}, [F(x)] \in [\underline{f}(x), \overline{f}(x)] \subset \mathbb{IR} \tag{4.2}$$

where $\underline{f}(x) \leq \overline{f}(x), \forall x \in Domain$.

Constrained interval functions (CIFs) are a generalization of constrained interval that are applied to interval-valued functions. They are generalized uncertainty representations albeit from a constraint interval point of view.

Definition 140 A **constrained interval function** (CIF), $F(x, \lambda_f)$, is defined for interval functions by

$$F(x, \lambda_f) = \underline{f}(x) + \lambda_f w_f(x), \tag{4.3}$$

where $w_f(x) = \overline{f}(x) - \underline{f}(x), \lambda_f \in [0, 1], x \in Domain$.

Definition 141 Let $\circ \in \{+, -, \times, \div\}$. CIF arithmetic (CIFA) is given by

$$H(x, \lambda_f, \lambda_g) = F(x, \lambda_f) \circ G(x, \lambda_g)$$
$$= \left(\underline{f}(x) + \lambda_f w_f(x)\right) \circ \left(\underline{g}(x) + \lambda_g w_g(x)\right) \tag{4.4}$$
$$\forall \lambda_f, \lambda_g \in [0, 1].$$

Example 142 Let $\underline{f}(x) = -|x| - 1, \overline{f}(x) = |x| + 1$. This means we have F defined as follows.

$$F(x, \lambda_f) = \underline{f}(x) + \lambda_f(\overline{f}(x) - \underline{f}(x))$$
$$= -|x| + 2\lambda_f(|x| + 1).$$

Depicted below for $\lambda_f = \frac{3}{4}$, $F(x, \frac{3}{4}) = \frac{1}{2}|x| + \frac{1}{2}$ and for $\lambda_f = \frac{1}{4}$, $F(x, \frac{1}{4}) = -\frac{1}{2}|x| - \frac{1}{2}$.

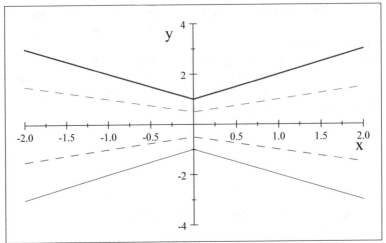

$$\text{Interval Functions} - \lambda = \tfrac{1}{4}, \tfrac{3}{4}$$

It is clear that with the CIF representation, generalized uncertainty analysis can be done from this point of view. Recall that interval-valued probabilities arise in the context of a cumulative distribution function $F_x(x)$ of a probability density function $p(x)$, when the cumulative $F_x(x)$ of $p(x)$ is only known to exist inside a bounding pair of functions $F_x(x) \in [\underline{F}(x), \overline{F}(x)]$. In this case, bounding possibility distributions (see [39] and Chap. 3) can be constructed such that

$$\{\Pr \mid \Pr \in [\underline{F}, \overline{F}]\} = \{\Pr(X \in A) \in [Pos_1(A), Pos_2(A)]\} \qquad (4.5)$$

for all measurable sets A which could have been derived independently of any fuzzy interval. Given the bounds, then the left and right distributions define the level sets and these can be are directly transformed into real-valued constraints. This was discussed in Chap. 3. Analysis of IVPs, since they are given by a bounding pair of functions, can be done from the CIF representation.

4.2.2 Soft Constraint Relationships

There are two approaches that lead to soft constraints: (1) The relationship $\leq$ or $=$ has a flexible meaning or right-hand side value is flexible (a range of values with preferences that encode, semantically, a *target* or a *goal*), (2) The constraint can accommodate "infeasibilities" up to a certain level. Both of these types are "gradual constraint set belonging" rather than what occurs in deterministic optimization which is a Boolean constraint set belonging. Here "constraint set belonging" is a matter of degree as a fuzzy set which is the first case. When the values tolerated on the right side are specified as in the second case, the degree of "feasibility" or "infeasibility" (tolerance) is specified as a possibility or necessity measure.

4.2.2.1   The Relationship or Right-Hand Side Value

A soft constraint relation is characterized by flexibility in what the relationship means. Relationships that are gradualness or flexible are fuzzy. For example, suppose we allow $\leq$ to mean that we are 100% satisfied with the inequality being non-negative and less than b_i but under no circumstances must we exceed $b_i + d_i$. Thus, one considers the most satisfying result from this flexible relationship as an x value belonging to the interval $[0, b_i + d_i]$. Additionally, we want the $y-$value of membership function to be 1 (completely satisfied) or at least as close as possible to 1 since this will mean that the constraint is either less than or equal to b_i or as close as possible to being less than or equal to b_i. If we have a right-hand side that is a trapezoidal fuzzy interval $0/0/b_i/b_i + d_i$, there are two ways to transform this fuzzy interval data into a real-valued constraint set. The first we call the Zimmermann approach since, to our knowledge, Zimmermann [13] was the first to use this. The second is via surprise functions of [40].

Zimmermann Approach

Case 143 The fuzzy interval right side is translated into a maximization of its $\alpha-$level real-valued equivalent as follows:

$$A_{io}x \tilde{\leq} b_i \text{ or } A_{io}x \leq \tilde{b}_i, i = 1, \dots, m$$

$$\updownarrow$$

$$z = \max \alpha$$
$$\alpha d_i + A_{io}x \leq b_i + d_i$$
$$0 \leq \alpha \leq 1,$$

where
$$A_{io}x = a_{i1}x_1 + \cdots + a_{in}x_n.$$

Recall that the tilde over the inequality means a fuzzy (soft) constraint relationship and over b_i means a fuzzy interval.

Case 144 As with constraint sets, a soft constraint relation is characterized by flexibility in what the relationship means. Relationships that are associated with transition or are flexible are fuzzy. For example, suppose we allow $\leq$ to mean that we are 100% satisfied with the inequality being less than z_* and non-negative, but under no circumstances exceed $z_* + d_0$. So one considers the most satisfying result from this flexible relationship first, as an x value belonging to the interval $[0, z_* + d_0]$ and secondly one whose corresponding membership function value, its $y-$value, is 1 or as close to 1 as possible, that is, an x in the interval $[0, z_*]$ for example. For body right-hand sides, assume that $\tilde{b} = 0/0/b/(b + d)$. Then the soft constraint is translated into a real-valued equivalent as follows:

$$\min f(c, x) = c^T x \le \tilde{z}_*$$
$$g(a, x) \le \tilde{b}$$
$$\updownarrow$$
$$\max \alpha$$
$$\alpha d_0 + c^T x \le z_* + d_0$$
$$\alpha d + g(a, x) \le b + d$$
$$0 \le \alpha \le 1,$$

where d_0 is the amount over z_* that would be tolerated as a minimum and d is the amount of flexibility over b that is possible.

If we have other types of fuzzy intervals, then the translations are as depicted in Figs. 4.1 and 4.2.

Surprise Function Approach

Surprise functions [40] are a newer approach for modeling soft constraints and it is a means of translating soft constraints and fuzzy interval parameter constraints into a function that acts as a *dynamic penalty* function when used in optimization. The derivation of soft constraints from classical or hard constraints is achieved as follows. Let

$$\text{hard } y_i = (A\vec{x})_i \le b_i \Rightarrow \text{soft } y_i = (A\vec{x})_i \le \tilde{b}_i, \tag{4.6}$$

where the right-hand side values of the soft constraint are *fuzzy intervals*. Let the fuzzy interval have membership function $\mu_i(x)$. A soft (fuzzy) inequality (4.6) with a fuzzy membership function, $\mu_i(x)$, is the possibility $Pos(\tilde{b}_i \ge Ax)$ and is translated into the surprise function

$$s_i(x) = \left(\frac{1}{\mu_i(x)} - 1\right)^2. \tag{4.7}$$

The surprise function (4.7) for a trapezoidal fuzzy interval is illustrated in Fig. 4.3. The way this is used in optimization is that we minimize the total surprise, $\sum_{i=1}^{m} s_i(x)$ where m is the total number of soft constraints. This is distinct from the Zimmermann approach [13].

Remark 145 Surprise functions, for our presentation, are restricted to fuzzy interval right-hand sides.

4.2.2.2 Tolerable Infeasibility (See [41])

The idea is to construct a way to allow infeasibilities to occur in such a way that they are greater than or equal to some minimal allowable infeasibility tolerance as measure by necessity or possibility, which are two of the usual ways to measure the degree of infeasibility. These are like chance constraints where the distributions are

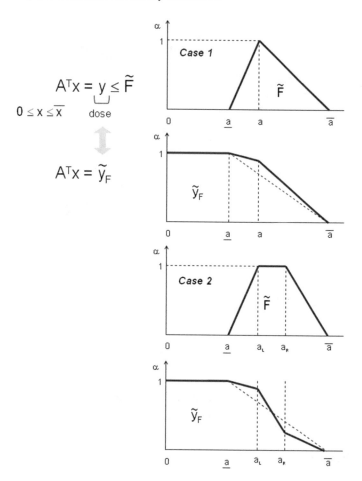

Fig. 4.1 Triangular and trapezoid

not probability distributions but possibilities or necessities. For constraints, we have (see [41]).

$$Nec(A_{io}x \leq b_i) \geq h_i^{nec},$$
$$Pos(A_{io}x \leq b_i) \geq h_i^{pos},$$
$$0 < h_i < 1.$$

For the objective function that has been transformed into a goal as done above, then we can use [41] as follows,

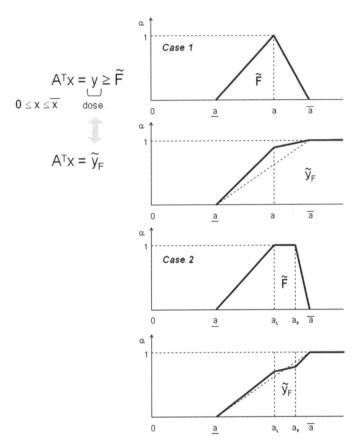

$$A^T x = y \geq \tilde{F}$$
$$\underbrace{}_{\text{dose}}$$
$$0 \leq x \leq \overline{x}$$

$$A^T x = \tilde{y}_F$$

Fig. 4.2 Greater than or equal constraint

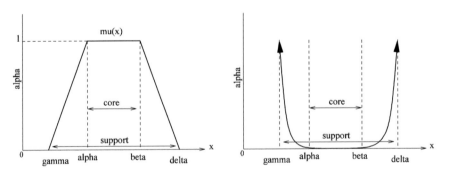

Fig. 4.3 Surprise function of a fuzzy interval

$$Nec(c^T x \leq z_*) \geq h,$$
$$Pos(c^T x \leq z_*) \geq h,$$
$$0 < h < 1,$$

where the meaning and real-valued equivalents of these inequalities will be derived later.

Example 146 Given a right-hand side value is fuzzy and we have a deterministic "equal to" for which the fuzzy interval right-hand side value represents a goal, for example "about 2" (see Fig. 1.4), we have a type of soft equality relationship that is translated into a real-valued equations as follows.

$$A_{io}x \leq \tilde{b}_i = \tilde{2} \tag{4.8}$$
$$\updownarrow$$
$$\alpha d_i^+ + A_{io}x \leq b_i + d_i^+$$
$$0 \leq \alpha \leq 1.$$

For our example of "about" 2,

$$\alpha + A_{io}x \leq 3, b_i = 2, d_i^+ = 1$$

where $\alpha = 1$ means that indeed the constraint is the real-number 2 and $\alpha = 0$ gives the limit to which our flexibility will be tolerated (a value between 1 and 3). In optimization, the alpha is the decision variable to be maximized. We note that this approach as developed by [13] in the context of fuzzy optimization is not guaranteed to be Pareto optimal. This is left as an exercise at the end of this chapter though there we give an example of this.

4.2.3 Soft Objective Relationships

Soft objective relationships can be thought of as goals. These occur in three broad categories.

1. "Traditional" goals, satisficing criteria such as "Come as close as possible to ...", "Maximize quality."
2. Multi-objective such as "Minimize investment risk and at the same time maximize investment profit."
3. Case-based such as "In all cases you must be feasible." or "Minimize the maximum regret."

4.3   Distinctions of Fuzzy and Possibility—Generalized Uncertainty in Optimization

We have mentioned before that there is often a confusion about the distinction between fuzzy or flexible and possibility or generalized uncertainty entities in general and specifically within the context of optimization. This section reconsiders this topic with the view of being clear and precise on how we, in this monograph, define and approach these distinctions within optimization. The greater important theoretical discussion about the difference between fuzzy and uncertainty we leave to others. Here, we try to garnish a few insights that will help us in our task of understanding how to construct and solve optimization models in which flexibilities and generalized uncertainty entities are a part of the problem and in some cases appear together within one optimization problem. First, however, a word about our distinction for this monograph between the words *uncertainty* and *fuzzy*.

4.3.1   Basic Notions

Remark 147 *Recall that we define **uncertainty** to be a one to many mapping.* Thus, a fuzzy set characterized by a membership function is **not** uncertainty according to how we define uncertainty since a fuzzy set is defined to be a (unique) single-valued membership function and not a one to many mapping. Moreover, we again emphasize that a set of possibilities defined by a pair of bounding functions is one type of generalized uncertainty so that possibilistic optimization is one type of optimization under generalized uncertainty.

The semantic distinctions between fuzzy and possibilistic optimization can also be found in [36, 42, 43] where it is noted that, for optimization models, *ambiguity* in the coefficients of the model leads to possibility optimization, while *vagueness* in the decision maker's preference is modeled by fuzzy optimization. When this vagueness represents a willingness on the part of the decision maker to relax his or her requirements in order to attain more suitable results, this type of fuzzy optimization we call *flexible programming*. It has also been said that, in the context of optimization, possibilistic uncertainty is *information-based* and flexible uncertainty is *preference-based* [22]. Another point of view on the semantic distinction between fuzzy or flexible and possibilistic uncertainty in optimization is articulated in Inuiguchi [42] who states:

> The membership grade of a fuzzy goal (fuzzy constraint) represents the *degree of satisfaction*, whereas that of a possibility distribution represents the *degree of occurrence*.

The semantics of an optimization problem are also influenced by where, structurally, in the optimization problem the fuzzy interval coefficient occurs. Suppose our optimization problem is known to have flexibility in the (in)equality constraints. In the case of soft constraints, it is the (in)equality itself that is viewed as fuzzy (i.e.

$Ax \widetilde{\leq} b$ for a linear programming problem). This is distinct from the case in which the right hand side has a vague value, in which case the right hand side is viewed to be fuzzy (i.e. $Ax \leq \tilde{b}$). To illustrate the difference between fuzzy inequalities and fuzzy right hand sides, Dr. Untiedt-Stock [3] considers the following.

Example 148 (*Computer dating service*) Suppose woman A specifies that she would like to date a "medium-height" man. Woman A defines "medium" as a fuzzy set characterized by a triangular membership function, centered at 69 in., with a spread of 3 in. This is a hard constraint because the man is *required* to be medium-height, albeit, "medium-height" is a fuzzy set. This is, therefore, an example of a fuzzy right-hand side (i.e. $Ax = \hat{b}$). Now consider woman B, who desires to date a man of 69 in. Unlike woman A, woman B is willing to compromise a little on the height requirement in order to be matched with a date who meets some of her other requirements. Her satisfaction level with a 69 in. man is 1, with a 68 or 70 in. man is $\frac{2}{3}$, and with a 67 or 71 in. man is $\frac{1}{3}$. This is an example of a soft constraint, represented by a fuzzy equality (i.e. $Ax \widetilde{=} b$). Notice that the membership functions in the two fuzzy cases are the same (symmetric triangular centered at 69 in. with a spread of 3 in.), but the semantics are different.

However, there is a further challenge, as we have already mentioned, not present in deterministic or stochastic optimization. In the context of fuzzy interval *coefficients*, a fuzzy right hand side that is used to encode the flexibility, as might occur for woman B is structurally indistinguishable from the requirement that woman A has in encoding her preference as the same fuzzy interval. When this is the case, it needs to be clearly specified how the right hand side fuzzy interval value b is being used semantically. In this text, we remind the reader that notationally, $\tilde{b}$ will denote a *fuzzy* semantic whereas $\hat{b}$ will denote a *generalized uncertainty* semantic and *both* are encoded by a fuzzy interval so that care must be taken in understanding the specific semantic of the problem. In particular, a right side that is fuzzy, $\tilde{b}$, is translated into $Ax \widetilde{=} b$ (see Figs. 4.1 and 4.2), or $\hat{b}$. which can be changed into a surprise, in the case of our example above.

Let us look at what we mean by fuzzy and possibility-generalized uncertainty, in the context of optimization, which we summarize in this following table.

	Fuzzy Set (Interval)	Possibility-Generalized Uncertainty
1. Representation	Single -valued function	Possibility/necessity or bounding pairs
2. Semantics	Gradualness	Information deficiency
3. Algebraic space	Set of real numbers, $\mathbb{R}$	Set of real functions
4. Analysis	Fuzzy interval analysis	Analysis of functions

Furthermore, Dr. Stock (see [3] p. 14) has a useful classification where we have changed possibilistic to generalized uncertainty and amplified the classification a bit.

Flexible/Fuzzy Optimization	Generalized Uncertainty Optimization
fuzzy - gradual belonging	imprecise
membership function	(possibility/necessity) distribution(s)
preference-based	information-based
vague	ambiguous
degree of satisfaction	degree of occurrence
aspirations/fuzzy targets	information deficiency

In the context of optimization there are some particular characteristics associated with our distinctions above. Fuzzy optimization models problems that contain *aspirations/goals* with no crisp set (Boolean) delineation are flexible. Generalized uncertainty optimization models problems that contain *ambiguities* arising from *incomplete/partial information* lead to optimization over functions. Thus, what must be clearly understood, before we undertake optimization, is that flexible and generalized uncertainty are distinct both in terms of their definitions and in terms of their meanings/semantics. At the same time, the way we encode or represent these two different entities are the same—a fuzzy interval. The constraint sets for flexible and generalized uncertainty and optimization methods are distinct because in one case we generate a constraint set of real numbers and in the other we generate a constraint set of real functions. This is quite different.

4.3.2 Relations: Flexible Versus Generalized Uncertainty in Constraint Sets

Let us distinguish types fuzzy interval number relationships that arise in constraints associated with optimization. There are more types of relationships than the three we present, but these three types are the generic types. The general real-valued (crisp) constraint relationship type is of the form

$$ax \ R \ by \qquad (4.9)$$

where $R \in \{\leq, =, \geq, \in\}$ with a, b real vector input coefficients (given or known *parameters*) and x, y real variable vectors where the context will make it clear what product is being used. This includes the case, for example,

$$g(a, x) \leq b$$

where we need to resolve a function of input vector coefficients. The linear case (4.9) is the generic one. In (4.9), we may have $y \equiv 1$. However, for this monograph, for generality, we can view the variable vector y as being a set in which case we would would have a scalar product and so accommodate the relationship $\in$ within this notation.

1. The relationship itself is fuzzy, that is,

$$ax \; \tilde{R} \; by \tag{4.10}$$

where x and y are real variable vectors and a and b real number parameter (input data) vectors and $\tilde{R}$ is a fuzzy relationship. For optimization, we have four main types of fuzzy relationships: $\tilde{\leq}, \tilde{=}, \tilde{\geq}, \tilde{\in}$.

2. The input parameters are fuzzy intervals, that is,

$$\tilde{a}x \; R \; \tilde{b}y \tag{4.11}$$

where x and y are real variable vectors and $\tilde{a}$ and $\tilde{b}$ are real fuzzy interval vector parameters (given/known input data) and R is a crisp relationship. For optimization, we have four main real-valued (crisp) relationships R: $\leq, =, \geq, \in$.

3. The right-hand side is a fuzzy interval, that is,

$$ax \; R \; \tilde{b} \tag{4.12}$$

where a is a real number parameter (input data) vector, x is a real vector and $\tilde{b}$ is a real parameter (given/known input data) fuzzy interval. In this case, depending of the semantics, we have a fuzzy relation, that is, flexibility, or a lack of information. If $R \; \tilde{b}$ represents a goal or aspiration, then (4.12) is flexibility. If $\tilde{b}$ arises from lack of information then we have a generalized uncertainty and we will denote this case by $\hat{b}$ to distinguish it from the flexible case, $\tilde{b}$.

The *first relationship* (4.10) will have the semantics of flexibility since we have from (4.10) with $\tilde{R}$ being $\tilde{\leq}$, "ax is nearly less than or equal to bx" and we can also substitute "about" and so on for "nearly". Thus relation (4.10) is a goal or an aspiration. That is, "It is our goal to come as close as possible to $ax \; R \; by$."

The *second relationship* (4.11) has parameters a and/or b modeled by real fuzzy interval number vectors where here $\tilde{a}$ is always a fuzzy interval vector. When we multiply the variables x and/or y by a fuzzy interval, this results in a generalized uncertainty or possibilistic distribution on the left and/or as on the right given the lack of information necessary to obtain a and b as real numbers. So we will consider (4.11) throughout the monograph as being part of generalized uncertainty. That is, *body coefficients that are fuzzy intervals generate generalized uncertainty constraints and hence generalized uncertainty optimization problems.*

The *third relationship* (4.12), for example, in radiation therapy, might have the right-hand side value for a tumor constraint inequality as a goal represented by a trapezoidal fuzzy interval 58/59/61/62. This means that a radiation oncologist would be 100% satisfied if the delivered dose at a tumor pixel were between 59 and 61 units. The radiation oncologist would be 100% dissatisfied if the tumor dose were below 58 or above 62. For dosages at the tumor pixel in the interval [58, 59] and [61, 62], the oncologist might have a linear preference, hence the trapezoidal fuzzy interval. This is an aspiration with preferences. On the other hand, if the trapezoidal fuzzy interval number 58/59/61/62 represents the partial information we have coming from research and oncologists' knowledge about the minimum dose value that will kill a cancerous tumor cell, then the trapezoidal number results in a possibility. That is, it is inconceivable to have a real number which *is* the *minimal dose that will kill a cancer cell*. One can model this via a probability distribution but inherently, the minimum dose of radiation dose that will kill a cancer tumor is in fact partial information based on research and experience given the impossibility to obtain such a value as a real number. When the fuzzy interval right-hand side represents a goal or aspiration, $\tilde{b}$, we translate it into a fuzzy (flexible) relationship which we will be explicit below. When the fuzzy interval right-hand side arises from deficiency of information, $\hat{b}$, we will translate these into surprise functions [40] or penalties as in [1]. Thus, we have two cases—fuzzy leading to a flexibility relation and generalized uncertainty leading to possibilistic relation. These are handled quite differently as we have intimated above and will delineate below. There are, of course, mixtures. However, for our exposition, we will develop the two given above and from these two types we are able to handle mixtures. In particular, we indicate how to handle mixtures via the two generic types of optimization (flexible and generalized uncertainty).

Let us look at the three relationships and how they, concretely, form constraint sets.

Example 149 Single Flexible/Fuzzy Aspiration: Consider the flexible/fuzzy aspiration/goal of being less than or equal to 2, that is $x \leq 2$. As an aspiration or flexibility in the context of fuzzy sets we have symbolically,

$$x \,\tilde{\leq}\, 2 \tag{4.13}$$

which might be turned into a membership function modeled as follows:

$$\mu_{x\tilde{\leq}2}(x) = \begin{cases} 1 \text{ if } x \leq 2 \\ -x + 3 \text{ if } 2 < x \leq 3 \\ 0 \text{ if } x > 3 \end{cases} . \tag{4.14}$$

Example 150 Single Generalized Uncertainty Relationship: Consider an ambiguous entity resulting from incomplete information in the right-hand side value 2 of a crisp $\leq$

$$x \leq \tilde{2} \tag{4.15}$$

where we're given $\tilde{2} = 2/2/3$, a triangular fuzzy number, so that in this case we have

$$(2/2/3) - x \geq 0$$

which is an ambiguous relationship. This results is another triangular fuzzy number relationship

$$(2 - x)/(2 - x)/(3 - x) \geq 0. \tag{4.16}$$

Remark 151 Note that (4.13) yields a **fuzzy membership function** (4.14) whereas (4.15) yields a **fuzzy relationship** (4.16). Given a particular instantiation of x for both cases, say, $x = \frac{5}{2}$, the fuzzy case (4.13) yields a **real number** $\frac{1}{2}$ whereas in the possibilistic case (4.15) yields the **fuzzy relationship** $-\frac{1}{2}/ - \frac{1}{2}/\frac{1}{2} \geq 0$.

A general constraint set has many such relations that must be satisfied simultaneously ("and-ed"). To "and" the first case (fuzzy, flexible) involves an **aggregation operator** when they occur as constraints. The second case is ambiguous and involves how to value simultaneously a set of functions like $(2 - x)/(2 - x)/(3 - x)$ with respect to being greater than or equal to a real number like 0 in our example. One method to value a set of simultaneously held functions with respect to a crisp relationship and a crisp right-hand side value is to apply a vector-valued functional to the vector of functions each of which has been obtained from the data just like (4.16) was obtained.

We next present the case of a single two-variable constraint set.

Example 152 Vector Flexible/Fuzzy Aspiration in a Single Relationship : Suppose we have a flexible constraint

$$2x_1 + 3x_2 \tilde{\leq} 5 \tag{4.17}$$

where $\tilde{\leq}5$ is given as

$$\mu_{\tilde{\leq}5}(x) = \begin{cases} 1 \text{ if } 2x_1 + 3x_2 \leq 5 \\ -(2x_1 + 3x_2) + 6 \text{ if } 5 < 2x_1 + 3x_2 \leq 6 \\ 0 \text{ if } 2x_1 + 3x_2 > 6 \end{cases} \tag{4.18}$$

Given and instantiation of $x = (x_1, x_2) = (\frac{5}{4}, 1)$,

$$\mu_{\tilde{\leq}5}(x) = \frac{1}{2} \tag{4.19}$$

which is a real number.

Example 153 Vector Generalized Uncertainty in a Single Relationship: Consider the possibilistic relationship

$$\tilde{2}x_1 + \tilde{3}x_2 \leq 5, \tag{4.20}$$

where

$$\tilde{2} = 1/2/3,$$
$$\tilde{3} = 2/3/4.$$

In this case, for $x_1, x_2 \geq 0$, we have

$$(1/2/3)x_1 + (2/3/4)x_2 \le 5, \tag{4.21}$$

$$(x_1 + 2x_2)/(2x_1 + 3x_2)/(3x_1 + 4x_2) \le 5 \tag{4.22}$$

Given the instantiation $x = (x_1, x_2) = (\frac{5}{4}, 1)$, we have

$$(1/2/3)\frac{5}{4} + (2/3/4)1 \le 5$$

$$\left(\frac{5}{4}/\frac{5}{2}/\frac{15}{4}\right) + (2/3/4) \le 5$$

$$\left(\frac{13}{4}/\frac{11}{2}/\frac{31}{4}\right) \le 5 \tag{4.23}$$

and we have a fuzzy interval related by a crisp $\le$ to a real-valued right-hand side.

Remark 154 Note the differences between the fuzzy/flexible and the generalized uncertainty cases. In one case we have a membership function(s) being the result of the relation and in the other, distributions (fuzzy intervals) in relationship result. In general, if we have n-variables and a single constraint, for the fuzzy case, we have a fuzzy membership function (4.18), which when instantiated yields a real number (4.19). In the generalized uncertainty case with n-variables and a single constraint, we have a fuzzy interval function on the left-hand side related via a crisp relationship to a real-valued right-hand side (4.21) which when instantiated yields a real fuzzy interval related via a crisp relation to a real-valued right hand side (4.23).

Simultaneous vector-valued equations will generate simultaneous fuzzy membership functions for the fuzzy/flexible case and simultaneous vector functions related via crisp relationships to real-valued vectors. This is illustrated in the following examples.

Example 155 System of Flexible/Fuzzy Aspiration: Now suppose we have two flexible constraint relations for $x_1 \ge 0, x_2 \ge 0$,

$$2x_1 + 3x_2 \tilde{\le} 5,$$

$$x_1 + 2x_2 \tilde{\le} 3, \tag{4.24}$$

where,

$$\mu_{\tilde{\le}5}(x) = \begin{cases} 1 \text{ if } 2x_1 + 3x_2 \le 5 \\ -(2x_1 + 3x_2) + 6 \text{ if } 5 < 2x_1 + 3x_2 \le 6 \\ 0 \text{ if } 2x_1 + 3x_2 > 6 \end{cases} \tag{4.25}$$

$$\mu_{\tilde{\le}3}(x) = \begin{cases} 1 \text{ if } x_1 + 2x_2 \le 3 \\ -(x_1 + 2x_2) + 3 \text{ if } 3 < x_1 + 2x_2 \le 4 \\ 0 \text{ if } x_1 + 2x_2 > 4 \end{cases} \tag{4.26}$$

For these simultaneous equations, we have

$$\mu_{\tilde{\leq}5}(x) = \begin{cases} 1 \text{ if } 2x_1 + 3x_2 \leq 5 \\ -(2x_1 + 3x_2) + 6 \text{ if } 5 < 2x_1 + 3x_2 \leq 6 \\ 0 \text{ if } 2x_1 + 3x_2 > 6 \end{cases}$$

and   (4.27)

$$\mu_{\tilde{\leq}3}(x) = \begin{cases} 1 \text{ if } x_1 + 2x_2 \leq 3 \\ -(x_1 + 2x_2) + 3 \text{ if } 3 < x_1 + 2x_2 \leq 4 \\ 0 \text{ if } x_1 + 2x_2 > 4 \end{cases}.$$

To resolve this system, the "and" must be concretized as an aggregation operator which is often the "min" (fuzzy intersection) operator.

Example 156 System of Generalized Uncertainty Relationships: Suppose we are given two possibilistic constraints for $x_1 \geq 0$, $x_2 \geq 0$,

$$\tilde{2}x_1 + \tilde{3}x_2 \leq 5,$$
$$\tilde{1}x_1 + \tilde{2}x_2 \leq 3,$$ (4.28)

where

$$\tilde{1} = 0/1/2,$$
$$\tilde{2} = 1/2/3,$$
$$\tilde{3} = 2/3/4.$$

This simultaneous generalized uncertainty system, results in

$$(1/2/3)x_1 + (2/3/4)x_2 \leq 5$$
$$(0/1/2)x_1 + (1/2/3)x_2 \leq 3$$

or multiplying a fuzzy interval number by $x_1 \geq 0$, $x_2 \geq 0$, and adding the fuzzy numbers yields

$$(x_1 + 2x_2)/(2x_1 + 3x_2)/(3x_1 + 4x_2) \leq 5$$
$$(0x_1 + x_2)/(x_1 + 2x_2)/(2x_1 + 3x_2) \leq 3.$$ (4.29)

Thus, to solve the simultaneous system (4.29), we must interpret what the meaning of two simultaneous relationships, which have fuzzy interval distributions on the left, a crisp relationship $\leq$, and a real number on the right.

Remark 157 In the case of a system of flexibility constraints, we have an aggregation membership functions whose instantiation results in a **real vector**. In the generalized uncertainty or possibilistic case, we have a system of functional relationships whose instantiation results in a **real-valued vector function relationship**, that is, on the left side we have a real-valued vector function crisply related to a real-valued vector right side. These, (4.24) and (4.28), are simultaneous so they are "and-ed". In the flexibility, fuzzy case, "and" is the t-norm (intersection of fuzzy sets, minimum of membership functions for example) which can be accomplished using one of the many **aggregation** operators. For the generalized uncertainty or possibilistic case, there are three broad approaches to "and" which we discuss in more detail later in the monograph. (1) The first is to impose a **scalarized functional** (a functional which maps a vector of functions onto the real numbers—see [44, 45]) on each left side of the constraint relationships (inequalities in our example above). For example, a necessity measure of whole relationship, called a modal value, is an example of this. That is, $Nec(\tilde{A}x \leq b)$ (see [46])is a modal value that can be considered as a scalarized functional (see [44, 45]). (2) The second approach is via a generalized uncertainty type of stochastic recourse approach where the constraints are moved directly into the objective function resulting in a penalty in objective function (to be minimized), where one such penalty functional is a generalized expected value called an **expected average** (see [47]). (3) The third approach considers the risk in taking one decision as opposed to another to obtain the most optimistic decision and the most pessimistic decision. In this case, a minimax regret might be imposed (or a fuzzy robust optimization approach).

4.3.3 Generalized Uncertainty in Constraint Sets

Optimization models usually involve constraints consisting of equalities, inequalities, or both. For deterministic optimization, the meaning and computation of the constraint set is clear. In flexible and/or generalized uncertainty optimization, however, the meaning and computation of "equality" and "inequality" as well as a set of simultaneous relations must be determined. To this end, Dubois and Prade, [48], gives a comprehensive analysis of fuzzy relations with four possible interpretations of fuzzy equalities. There are two broad approaches to resolving inequalities and equalities. The first is to go $\alpha-$level by $\alpha-$level. The second is to use necessity and/or possibility measures to indicate the minimum infeasibility tolerated (where we seek to do better— maximize feasibility as much as possible but never to go below the specified measure). In stochastic optimization, these are chance-constraints where the probability of infeasibility is used as the measure limiting the minimum probability tolerated. In generalized uncertainty, these are called *modalities* and unlike probability where there is one measure, in generalized uncertainty, there are six types of modalities. The $\alpha-$level by $\alpha-$level approach has four main types.

4.3.3.1 Alpha-Level by Alpha-Level

Let us look at the constraint set as a continuum of $\alpha-$levels of real-valued interval relationships. That is, equations like the generalized uncertainty inequality relationship given by Eq. (4.21), each $\alpha-$level for the corresponding fuzzy sets associated with the left side of an equation like (4.21), say, $\tilde{M}$ and right side of an equation like (4.21), say, $\tilde{N}$ (though in the example of (4.21), the right side is a real number which, is also a fuzzy interval represented by a characteristic function, but the right side could be a fuzzy interval as well), is given by $\tilde{M}(\alpha) = [m^-(\alpha), m^+(\alpha)]$ and $\tilde{N}(\alpha) = [n^-(\alpha), n^+(\alpha)]$, respectively. There are four main types of fuzzy interval relationships that can be thought of as the most basic and have their origin in interval analysis (see [49]). The statement $\tilde{M} \geq \tilde{N}$ can be interpreted in any of the four following ways where we drop the α since they are all α−levels:

1. $\forall x \in \tilde{M}, \forall y \in \tilde{N}, x > y$. This is equivalent to $m^- > n^+$;
2. $\forall x \in \tilde{M}, \exists y \in \tilde{N}, x \geq y$. This is equivalent to $m^- \geq n^-$;
3. $\exists x \in \tilde{M}, \forall y \in \tilde{N}, x > y$. This is equivalent to $m^+ > n^+$;
4. $\exists (x, y) \in \tilde{M} \times \tilde{N}, x \geq y$. This is equivalent to $m^+ \geq n^-$.

Inequality relation (1) indicates that x is necessarily greater than y, This is the pessimistic view. The decision maker who requires that $m^- > n+$ in order to satisfy $\tilde{M} > \tilde{N}$ is taking no chances [50]. (4) indicates that x is possibly greater than y. This is the optimistic view. The decision maker who merely requires that $m^+ > n^-$ in order to satisfy $\tilde{M} > \tilde{N}$ has a hopeful outlook. Inequality relations (2) and (3) fall somewhere between the optimistic and pessimistic views (see the classification by Lodwick [50]).

The statement $\tilde{M} = \tilde{N}$ can be interpreted in any of the following four ways (also see [49]):

1. Zadeh's fuzzy set equality: $\mu_M = \mu_N$
2. $\forall x \in \tilde{M}(\alpha), \exists y \in \tilde{N}(\alpha), x = y$ for all $0 \leq \alpha \leq 1$ (which is equivalent to $\tilde{M}(\alpha) \subseteq \tilde{N}(\alpha)$).
3. $\forall y \in \tilde{N}(\alpha), \exists x \in \tilde{M}(\alpha), x = y$ for all $0 \leq \alpha \leq 1$ (which is equivalent to $\tilde{N}(\alpha) \subseteq \tilde{M}(\alpha)$).
4. $\exists (x, y) \in \tilde{M}(\alpha) \times \tilde{N}(\alpha), x = y$ for all $0 \leq \alpha \leq 1$ (which is equivalent to $\tilde{N}(\alpha) \cap \tilde{M}(\alpha) \neq \emptyset$.)

Equality relation (1) indicates that x is necessarily equal to y (the pessimistic view), (4) indicates that x is possibly equal to y (the optimistic view), and (2) and (3) fall somewhere in between.

4.3.3.2 Modalities

Chance-type constraints arise is generalized uncertainty optimization constraint sets. Given that we have lack of information, as we have delineated in our previous chapters, necessity and possibility is a way of dealing with ambiguity in the left-hand side and/or the right-hand side coefficients. Thus, to resolve an ambiguity arising in a constraint relationship, for example, as in Eq. (4.21), one translates this type of equation into a necessity or possibility. That is, equation

$$\hat{a}_{i1}x_1 + \hat{a}_{i2}x_2 + \cdots + \hat{a}_{in}x_n \leq \hat{b}_i$$

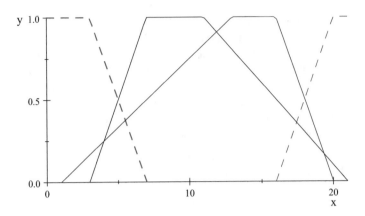

Fig. 4.4 Modal inequalities

where at least one of the coefficients a, b are fuzzy intervals whose semantic is tied to information deficiency (rather than set belonging in the case of b), then to resolve this, we need to find a measure. The chance constraint type of measure is the necessity or possibility measure which symbolically given as

$$Nec(\hat{a}_{i1}x_1 + \hat{a}_{i2}x_2 + \cdots + \hat{a}_{in}x_n \leq \hat{b}_i), \tag{4.30}$$

$$Pos(\hat{a}_{i1}x_1 + \hat{a}_{i2}x_2 + \cdots + \hat{a}_{in}x_n \leq \hat{b}_i). \tag{4.31}$$

Inuiguchi, Ichihashii, and Kume [43] present six modalities (also see [46]). What these symbols mean in terms of a real-valued equation will be discussed in Chap. 6. Here, it suffices to state that chance constraint equivalents in fuzzy constraints is always, for this manuscript, considered as part of generalized uncertainty optimization constraint sets and this is how we translate a constraint equation such as Eq. (4.21) into a real-valued equation. Ways to translate these types of equations are called modalities. We use two (4.30), (4.31). The dark solid lines of Fig. 4.4 represent the membership function $\mu_M(x)$ and the dark dotted line is $\mu_N(x)$. The solid light line is $1 - \mu_M(x)$ where the light dotted line is $1 - \mu_N(x)$. Point a is $Nec(x^- \geq y^-)$. Point b is $Pos(x^+ \geq y^-)$, point c is $Pos(x^+ > y^+)$ (see Fig. 4.4).

We translate flexible and generalized uncertainty constraints into real-valued constraint equivalents. There are three types that arise and we discuss this next.

4.3.4 Three Types of Constraints: Semantic Considerations

Considering semantics, we have three types of constraint sets. They are (1) pure flexible, (2) pure generalized uncertainty, and (3) mixed flexible and generalized uncertainty. Let us look at these one at a time.

1. **Pure flexible constraint sets**: Purely flexible constraints are fuzzy goals or aspirations, (4.13), (4.17), or (4.24), leading to (4.14), (4.18), (4.25), (4.26). That is, a flexible constraint set is of the type (4.26). The general way to translate this constraint set into a real-valued constraint set (a crisp constraint set) is to use the translation given by (4.33). These steps are the symbolic representation of the problem. Our examples make this concrete.
2. **Pure generalized uncertainty constraint sets**: Purely generalized constraints (4.29) are translated according to (4.34) which is a real-valued constraint set specification.
3. **Mixed flexible and generalized uncertainty**: Suppose every constraint relation (every equality, inequality or set belonging) is of only one type, either completely flexible or completely generalized uncertainty and we have at least one type of each in the set of relationships that constitute the constraint set. When we have a single type of relationship in the constraints, we handle the flexible ones via (4.33) and the generalized uncertainty ones via (4.34). That is, we use a fuzzy aggregation on the flexible constraints and one of the three types of functionals U (value, penalty, or minimax regret) on the generalized uncertainty constraints.

We make these observations more precise in what follows. However, let us complete our comparison of flexible and generalized uncertainty by looking at what happens in the objective function part of our optimization structure.

4.3.5 Objective Function: Flexible Versus Generalized Uncertainty in Objective Functions

This monograph focuses on two types of objective function cases.

1. The operator *opt* is fuzzy, that is, $z = \widetilde{opt}$.
2. The cost coefficient vector is c is a fuzzy interval vector, that is, $opt\ z = f(\tilde{c}, x)$.

We discuss these two cases in more detail.

4.3.5.1 Fuzzy Optimization Operator

A fuzzy optimization operator "optimize" means "come as close as possible" as we have previously discussed. In this case, given a real number target that concretizes what is meant by "come as close as possible", the objective becomes a flexible constraint. The objective becomes to find decision that will put us closest to the constraints.

4.3.5.2 Fuzzy Cost Coefficient

Fuzzy cost coefficients c of the objective function yields a fuzzy interval function relationship with a real-valued relationship $=$ and a real valued variable z to which the operator *opt* is applied. For example, for a linear two variable case, we might have,

$$z = \tilde{2}x_1 + \tilde{3}x_2.$$

If

$$\tilde{2} = 1/2/3,$$
$$\tilde{3} = 2/3/4,$$

and $x_1, x_2 \geq 0$, we have

$$opt\ z = (x_1 + 2x_2)/(2x_1 + 3x_2)/(3x_1 + 4x_2). \tag{4.32}$$

That is, we are optimizing a triangular fuzzy interval objective function. Note that for the objective function, we are optimizing a functional objective. The n-dimensional linear case is follows the same pattern as (4.32). Clearly, a way to map the right side of (4.32) onto the real numbers is needed. Optimal control problems use an integral objective criterion such as the minimization of energy (the integral being the measure of energy) to optimize over functions. There is also an "expected value" measure of function, which is also an integral. We will return to this point later in the text.

4.4 Solution Methods

Once a problem has been identified as flexible or generalized uncertainty, solutions methods for each are distinct. Next what is meant by decision-making in the presence of flexible and generalized uncertainty entities is defined (see [1]).

1. *Flexible Decision Making:* Given the set of real-valued (crisp) decisions, Ω, and fuzzy sets, $\{\tilde{F}_i \mid i = 1,\ldots, n\}$, find the optimal decision in the set Ω. That is,

$$\sup_{x \in \Omega} h\left(\tilde{F}_1(x), \ldots, \tilde{F}_n(x)\right), \tag{4.33}$$

where $h : [0, 1]^n \to [0, 1]$ is an *aggregation* operator [51] often taken to be the *min* operator, and $\tilde{F}_i(x) \in [0, 1]$ is the fuzzy membership of x in fuzzy set $\tilde{F}_i$. The decision space Ω is a set of real numbers (so called *crisp set*), and the optimal decision satisfies a mutual membership condition defined by the aggregation operator h. Note that this is precisely what [52] states in terms of a fuzzy set. The elements (in our case x) are distinct. The set belonging is flexible. This is the method of Bellman and Zadeh [8], Tanaka, Okuda and Asai [10, 11], and Zimmermann [13], who were the first (in this order) to develop fuzzy mathematical programming. While the aggregation operator h historically has been the *min* operator, it can be, for example, any $t - norm$ that is consistent with the context of the problem and/or decision methods (see [53] or [54]). If the traditional aggregation function h is not normalized to [0, 1] but allowed to be $\mathbb{R}^+ = [0, \infty)$, that is, $h : [0, 1]^n \to [0, \infty)$, then (4.33) explicitly includes surprise functions (4.7). In particular surprise functions (4.7) can also be used as aggregation operators.

2. *Generalized Uncertainty Decision Making:* Given the set of real-valued (crisp) decisions, Ω, and the set of generalized distributions representing the uncertain outcomes from selecting decision $\vec{x} = (x_1, \ldots, x_n)^T$ denoted $\Psi_x = \{\hat{F}_x^i, i = 1, \ldots, n\}$, find the optimal decision that produces the best set of possible outcomes with respect to an ordering U of the outcomes. That is,

$$\sup_{\Psi_x \in \Psi} U(\Psi_x), \qquad (4.34)$$

where $U(\Psi_x)$ represents an "evaluation **function**" (also called utility function, expectation, or scalarized function—[44, 45]) of the set of distributions of possible outcomes $\Psi = \{\Psi_x | x \in \Omega\}$. The decision space Ψ is a **set of generalized uncertainty distributions** $\Psi_x : \Omega \to [0, 1]$ resulting from taking decision $x \in \Omega$. This is the semantic taken in the possibility optimization [1, 36, 42, 43]. For example, if $\hat{F}_x = \hat{2}x_1 + \hat{3}x_2$, where $\hat{2}$ and $\hat{3}$ are the possibility numbers 2 and 3, then each $\vec{x} = (x_1, x_2)^T$ generates the possibility distribution $\hat{F}_x = \hat{2}x_1 + \hat{3}x_2$. Note that here $\hat{F}$ is well-defined. The degree to which a variable x belongs (or is) $\hat{F}$ is measured through the possibility distribution $\hat{F}_x$. This is what [52] observes in the case of possibility.

Remark 158 Let us summarize what we have just stated because we consider this an important point often forgotten. For fuzzy sets $\tilde{F}_i$, $i = 1, \ldots, n$, given x, $[\tilde{F}_1(x), \ldots, \tilde{F}_n(x)]^T$ is a real-valued **vector**. Thus, we need a way to aggregate the components of the vectors into a single real-value. This is done by a t-norm, min for example. For generalized uncertainty, given x, $\Psi_x = \{\hat{F}_x^i, i = 1, \ldots, n\}$ is a set of **distributions**, so we need a way to turn this set of distributions into a single real-value, which may be implemented using an evaluation function, a generalized expectation with recourse, for example. Very simply, fuzzy decision-making selects from a set of real-valued, crisp, elements ordered by an aggregation operator on corresponding membership functions, while generalized uncertainty decision making selects from a set of distributions measured by an evaluation operator that orders sets of distributions.

These two different approaches have two different ordering operators (an aggregation operation for fuzzy sets such as *min* and a scalarized function in the case of generalized uncertainty such as a generalized expectation—see [44, 45]) and lead to two different optimization methods. The underlying constraint sets associated with fuzzy decision-making are fuzzy, where one forms the decision space of real-valued elements from operations on these fuzzy sets ("*min*" and "*and*", for example, in the case of optimization of [8, 11, 13]). The underlying constraint sets associated with generalized uncertainty decision making are sets of real-valued distribution functions.

4.5 Summary

We summarize what has been articulated using, additionally, insights from [55].

- Fuzzy optimization (what we call here *flexible optimization*), offers a bridge between numerical (deterministic) approaches and the linguistic or qualitative ones. The thrust of flexible and generalized uncertainty approaches are to provide the analyst with a coherent way to distinguish between fuzzy and generalized uncertainty in decision processes.
- Fuzzy set theory and its mathematical environment of aggregation operators ("*and*", *t-norms*), interval analysis, constraint interval analysis [56, 57]), fuzzy interval analysis, constraint fuzzy interval analysis [56, 57], gradual numbers (see [58–60]), and preference modeling, provide a general framework for posing decision problems in a more open way and provides a unification of existing techniques and theories.
- Fuzzy set theory has the capacity of translating linguistic variables into quantitative terms in a flexible and useful way.
- Generalized uncertainty and possibility theory explicitly accounts for lack of information, avoiding the use of an unique, often considered as uniform, probability distributions.
- The set theoretic view of functions to represent numbers (intervals) on which scalarization functions (see [44, 45]) are expressed as fuzzy sets offer a wide range of aggregation operations and functionals.

In short, fuzzy set theory and possibility theory offer optimization an approach that comes close to the underlying flexible, satisficing, and information deficient processes that are very often the typical environment in which decision makers find themselves.

4.6 Exercises

We have the following linear programming problem:

$$\max z = x_1 + x_2 \tag{4.35}$$
$$2x_1 + 3x_2 \le 12$$
$$2x_1 + x_1 \le 8$$

$$x_1, x_2 \geq 0$$

Suppose for the constraints the flexibility, $d_1 = 4$, and $d_2 = 4$. From a violation of tolerances, suppose we can tolerate up to a 25% violation, that is, we want the constraints to be 75% feasible.

Exercises 159 Look up Zimmermann's fuzzy linear programming method (see [13] or Chap. 5) and write down the translated real-valued linear programming problem for (4.35) using Zimmermann's approach.

Exercises 160 Look up Inuiguchi, Ichihashi, Kume's possibilistic linear programming method (see [43] or Chap. 6) and write down the translated real-valued linear programming problem for (4.35) and the six cases of modal values according to Inuiguchi.

Exercises 161 Write down the translated real-valued linear programming problem for (4.35) using the surprise function approach.

Exercises 162 Compute the optimal solution of (4.35) using the Zimmermann approach.

Exercises 163 Compute the optimal solution of (4.35) using the surprise approach.

Exercises 164 Prove that the Zimmermann approach for (4.35) is not Pareto optimal.

Exercises 165 Prove that the surprise approach for (4.35) is Pareto optimal.

References

1. W.A. Lodwick, K.D. Jamison, Theory and semantics for fuzzy and possibility optimization. Fuzzy Sets Syst. **158**(17), 1861–1871 (2007)
2. W.A. Lodwick, M. Inuiguchi, Professors Kiyoji Asai's and Hideo Tanaka's 1973–1983: flexible and generalized uncertainty optimization approaches revisited. IFSA 2013, Invited Talk, June 25 (2013)
3. E. Untiedt, Fuzzy and possibilistic programming techniques in the radiation therapy problem: an implementation-bases analysis. Masters Thesis, University of Colorado Denver (2006)
4. H. Greenberg, Intelligent mathematical programming meeting, University of Colorado Denver (1988)
5. J. Ramik, J. Rimanek, Inequality relation between fuzzy numbers and its use in fuzzy optimization. Fuzzy Sets Syst. **16**, 123–138 (1985)
6. J. Ramik, M. Vlach, Fuzzy mathematical programming: a unified approach based on fuzzy relations. Fuzzy Opt. Decis. Making **1**, 335–346 (2002)
7. J. Ramik, M. Vlach, *Generalized Concavity in Fuzzy Optimization and Decision Analysis* (Kluwer Academic Publishers, Boston, 2002)
8. R.E. Bellman, L.A. Zadeh, Decision-making in a fuzzy environment. Manag. Sci. Ser. B **17**, 141–164 (1970)

9. K. Asai, H. Tanaka, On the fuzzy-mathematical programming, in *Proceedings of the 3rd IFAC Symposium on Identification and System Parameter Estimation*, The Hague/Delft, The Netherlands, 12–15 June 1973, pp. 1050–1051

10. H. Tanaka, T. Okuda, K. Asai, On fuzzy mathematical programming. Trans. Soc. Instr. Control Eng. **9**(5), 607–613 (1973). (in Japanese)

11. H. Tanaka, T. Okuda, K. Asai, On fuzzy mathematical programming. J. Cybern. **3**, 37–46 (1974)

12. H. Zimmermann, Optimization in fuzzy environment, in *Paper presented at the XXI TIMS and 46th ORSA Conference* (Puerto Rico Meeting, San Juan, Puerto Rico, 1974)

13. H. Zimmermann, Description and optimization of fuzzy systems. Int. J. Gen. Syst. **2**, 209–215 (1976)

14. C.V. Negoita, D.A. Ralescu, *Applications of Fuzzy Sets to Systems Analysis* (Birkhauser, Boston, 1975)

15. C.V. Negoita, N. Sularia, On fuzzy mathematical programming and tolerances in planning. Econ. Comput. Econ. Cybern. Stud. Res. **1**, 3–15 (1976)

16. W.A. Lodwick, M. Inuiguchi (ed.), Fuzzy and possibilistic optimization, *Fuzzy Sets and Systems* (2007)

17. M. Inuiguchi, W.A. Lodwick, Foundational contributions of K. Asai and H. Tanaka to fuzzy optimization. Fuzzy Sets Syst. **274**, 24–46 (2015)

18. M. Delgado, J. Kacprzyk, J.-L. Verdegay, M.A. Vila, *Fuzzy Optimization: Recent Advances* (Physica-Verlag, Heidelberg, 1994)

19. J. Kacprzyk, S.A. Orlovski, *Optimization Models Using Fuzzy Sets and Possibility Theory* (D. Reidel Publishing Company, Dordrecht, 1987)

20. W.A. Lodwick, J. Kacprzyk, *Fuzzy Optimization: Recent Developments and Applications* (Springer, New York, 2010)

21. C.R. Bector, S. Chandra, *Fuzzy Mathematical Programming and Fuzzy Matrix Games* (Springer, Berlin, 2005)

22. Y. Lai, C. Hwang, *Fuzzy Mathematical Programming* (Springer, Berlin, 1992)

23. B. Liu, *Uncertainty Programming* (Wiley, New York, 1999)

24. B. Liu, *Theory and Practice of Uncertainty Programming* (Physica-Verlag, Heidelberg, 2002)

25. J.M. Sousa, U. Kaymak, *Fuzzy Decision Making in Modeling and Control* (World Science Press, Singapore, 2002)

26. M.K. Luhandjula, On possibility linear programming. Fuzzy Sets Syst. **18**, 15–30 (1986)

27. M.K. Luhandjula, Optimization under hybrid uncertainty. Fuzzy Sets Syst. **146**, 187–203 (2004)

28. M.K. Luhandjula, Fuzzy stochastic linear programming: survey and future directions. Eur. J. Oper. Res. **174**, 1353–1367 (2006)

29. N. Sahinidis, Optimization under uncertainty: state-of-the-art and opportunities. Comput. Chem. Eng. **28**, 971–983 (2004)

30. J.J. Buckley, Possibility and necessity in optimization. Fuzzy Sets Syst. **25**(1), 1–13 (1988)

31. M. Inuiguchi, Fuzzy linear programming, what, why and how? Tatra Mountain Pub. **13**, 123–167 (1997)

32. M. Inuiguchi, H. Ichihashi, H. Tanaka, Fuzzy programming: a survey of recent developments, in *Stochastic versus Fuzzy Approaches to Multiobjective Mathematical Programming under Uncertainty*, ed. by R. Slowinski, J. Teghem (Kluwer Academic Publishers, Dordrecht, 1990), pp. 45–68

33. W.A. Lodwick, E. Untiedt, Chapter 1: fuzzy optimization, in *Fuzzy Optimization: Recent Developments and Applications*, eds. by W.A. Lodwick, J. Kacprzyk (Springer, New York, 2010)

34. J.L. Verdegay, Fuzzy mathematical programming, in *Fuzzy Information and Decision Processes*, eds. by M.M. Gupta, E. Sanchez (North Holland Company, Amsterdam, 1982), pp. 231–237

35. H. Zimmermann, Fuzzy mathematical programming. Comput. Oper. Res. **10**, 291–298 (1983)

36. M. Inuiguchi, M. Sakawa, Y. Kume, The usefulness of possibility programming in production planning problems. Int. J. Prod. Econ. **33**, 49–52 (1994)

37. W.A. Lodwick, R. Jafelice Motta, Constraint interval function analysis – theory and application to generalized expectation in optimization, in *Proceedings, NAFIPS 2018/CBSF V, Fortaleza, Ceará, Brazil, 4–6 July* (2018)

38. W.A. Lodwick, M.T. Mizukoshi, *Fuzzy Constraint Interval Mathematical Analysis – An Introduction*, NAFIP2020, Redmond, WA. Accessed 20–22 August 2020

39. K.D. Jamison, W.A. Lodwick, The construction of consistent possibility and necessity measures. Fuzzy Sets Syst. **132**, 1–10 (2002)

40. A. Neumaier, Fuzzy modeling in terms of surprise. Fuzzy Sets Syst. **135**(1), 21–38 (2003)

41. M. Inuiguchi, Robust optimization by means of fuzzy linear programming, in *Managing Safety of Heterogeneous Systems: Decisions under Uncertainties and Risks*, eds. by Y. Ermoliev, M. Makowski, K. Marti, LNEMS, vol. 658. (Springer, Berlin, 2012), pp. 219–239

42. M. Inuiguchi, Stochastic programming problems versus fuzzy mathematical programming problems. Jpn. J. Fuzzy Theory Syst. **4**(1), 97–109 (1992)

43. M. Inuiguchi, H. Ichihashi, Y. Kume, Relationships between modality constrained programming problems and various fuzzy mathematical programming problems. Fuzzy Sets Syst. **49**, 243–259 (1992)

44. M. Ehrgott, *Multicriteria Optimization* (Springer Science & Business, Berlin, 2006)

45. J. Jahn (ed.), *Vector Optimization* (Springer, Berlin, 2009)

46. M. Inuiguchi, Necessity measure optimization in linear programming problems with fuzzy polytopes. Fuzzy Sets Syst. **158**, 1882–1891 (2007)

47. K.D. Jamison, W.A. Lodwick, Fuzzy linear programming using penalty method. Fuzzy Sets Syst. **119**, 97–110 (2001)

48. D. Dubois, H. Prade, Fuzzy numbers: an overview, in *Analysis of Fuzzy Information, Volume I: Mathematics and Logic*, ed. by J.C. Bezdek. (CRC Press, 1987), pp. 3–39

49. W.A. Lodwick, D. Dubois, Interval linear systems as a necessary step in fuzzy linear systems. Fuzzy Sets Syst. **274**, 227–251 (2015)

50. W.A. Lodwick, Analysis of Structure in Fuzzy Linear Programs. Fuzzy Sets Syst. **38**(1), 15–26 (1990)

51. G.J. Klir, B. Yuan, *Fuzzy Sets and Fuzzy Logic: Theory and Applications* (Prentice Hall, New Jersey, 1995)

52. D. Dubois, H. Prade, Formal representations of uncertainty (Chapter 3), in *Decision-Making Process, ISTE, London, UK*, eds. by D. Bouyssou, D. Dubois, H. Prade. (Wiley, Hoboken, 2009)

53. U. Kaymak, J.M. Sousa, Weighting of constraints in fuzzy optimization. Constraints **8**, 61–78 (2003)

54. H.J. Rommelfanger, T. Keresztfalvi, Multicriteria fuzzy optimization based on Yager's parameterized t-norm. Found. Comput. Decis. Sci. **16**, 99–110 (1991)

55. D. Dubois, The role of fuzzy sets in decision sciences: old techniques and new directions. Fuzzy Sets Syst. **184**(1), 3–28 (2010)

56. W.A. Lodwick, Constrained interval arithmetic. CCM Report 138 (1999)

57. W.A. Lodwick, Fundamentals of interval analysis and linkages to fuzzy set theory, in *Handbook of Granular Computing*, eds. by W. Pedrycz, A. Skowron, V. Kreinovich (Wiley, West Sussex, 2008), pp. 55–79

58. J. Fortin, D. Dubois, H. Fargier, Gradual numbers and their application to fuzzy interval analysis. IEEE Trans. Fuzzy Syst. **16**, 388–402 (2008)

59. E. Stock, Gradual numbers and fuzzy optimization. Ph.D. Thesis, University of Colorado Denver (2010)

60. E. Untiedt, W.A. Lodwick, On selecting an algorithm for fuzzy optimization, in *Foundations of Fuzzy Logic and Soft Computing: 12th International Fuzzy System Association World Congress, IFSA 2007, Cancun, Mexico, June 2007, Proceedings*, eds. by P. Melin, O. Castillo, L.T. Aguilar, J. Kacpzryk, W. Pedrycz. Springer, pp. 371–380

Chapter 5
Flexible Optimization

5.1 Introduction

We have three generic flexible optimization types.

1. Optimization types arising from the original approach to fuzzy optimization introduced in 1970 by Bellman and Zadeh [1] which they called "decision making in a fuzzy environment" and is the basis for most flexible optimization approaches. For these, there are two ways to optimize:

   (a) The Tanaka, Okuda, and Asai [2] approach;
   (b) The Zimmermann [3] approach.

2. Optimization types arising from fuzzy interval solutions based on the work of Verdegay, Delgado, Vila, Kacprzyk [4–6].
3. Optimization types arising from fuzzy interval right-hand sides using surprise functions of Neumaier [7] as developed by Lodwick, Newman, and Neumaier [8].

These are the themes of this chapter.

Remark 166 Much of the literature uses the ter "fuzzy optimization" for optimization problems with fuzzy interval left-sides coefficients in the body and resulting fuzzy interval inequality relations. As we have mentioned, this monograph maintains that the semantics based on fuzzy parameters arising from the coefficients delineating the body of the optimization model most properly falls within the generalized uncertainty category. This case of fuzzy interval coefficients that occur in the body of the model is discussed in the next chapter. We note that this case of fuzzy interval left-side body coefficients means that the model, as specified, is ambiguous. That is, the model as specified arises from incomplete or partial information. As we have mentioned before, this monograph clearly separates what is optimization over transitional entities and optimization over generalized uncertainty entities using the term flexible optimization for the former, rather than fuzzy optimization, and gen-

© Springer Nature Switzerland AG 2021
W. A. Lodwick and L. L. Salles-Neto, *Flexible and Generalized
Uncertainty Optimization*, Studies in Computational Intelligence 696,
https://doi.org/10.1007/978-3-030-61180-4_5

eralized uncertainty optimization for the latter, rather than possibilistic optimization since generalized uncertainty is broader than possibility.

5.2   Flexible Optimization

Flexible optimization, as we have mentioned, is optimization associated with fuzzy goals or fuzzy aspirations. As such, flexible optimization optimizes set membership in a fuzzy set. The fuzzy set over which we optimize is the result of turning the fuzzy objective function into a fuzzy constraint and adding it to the fuzzy relationships in the constraint set. The fuzzy constraints are then transformed into real-valued matrix-vector constraints by a fuzzy aggregation operator. Thus, our approach is to:

1. *First* change all fuzzy goals into fuzzy relational constraints.
2. *Second*, given that we have fuzzy relational constraints, we "and" them, that is, we intersect them and this is our fuzzy constraint set. In the context of fuzzy set theory, as we have mentioned before, this means an *aggregation* operator and there are various aggregation/intersection operators, the usual ones are the t-norms.
3. *Third,* optimize, that is, find the highest possible resulting membership value

Remark 167 The deterministic case has only one interpretation of intersection in the context of constraint sets and that is finding simultaneous solutions to the relations which is the classical set intersection. That is, simultaneous solutions are the intersection of solution sets of the individual relations of the system of constraints, thus, there is only one way to deterministically intersect sets in $\mathbb{R}^n$. While any t-norm can be used in aggregating to obtain the resultant fuzzy set associated with systems of fuzzy relations, our exposition just uses "min" which may be thought of as the simplest within the suite of t-norms where generalizations to other t-norms is straight forward and thus not presented here.

The general flexible optimization problem is:

$$\widetilde{opt}_{\substack{x}} \; z = f(\vec{c}, \vec{x}) \tag{5.1}$$

$$\vec{x} \, \tilde{\in} \, \tilde{X} \tag{5.2}$$

where we drop the vector notation and where we obtain $x \, \tilde{\in} \, \tilde{X}$ from

$$g(a, x) \, \tilde{R} \, b \tag{5.3}$$

and $\tilde{R}$ is a fuzzy relationship. The following steps are useful in delineating a flexible optimization problem.

1. Identify the *fuzzy components* that make up the flexible optimization model:

   (a) Objective function as a target or goal;

(b) Fuzzy relationship in the constraint definition where the constraint is an aspiration (target with flexibility);

(c) Fuzzy interval right-hand side value that has a gradual set belonging semantic;

2. *Translation* of components into fuzzy sets, that is, compute the associated fuzzy membership function where it occurs (objective function and/or fuzzy relationship and/or fuzzy interval right-hand side)

3. *Transformation* into a real-valued mathematical programming model.

The general flexible optimization problem is to find a fuzzy optimum of a fuzzy objective function subject to fuzzy constraints. The variables are the unknown quantities whose values are to be determined by the model.

5.2.1 Fuzzy Components of a Flexible Optimization Model

The first fuzzy component that we construct is associated with the objective function. The second we construct is a fuzzy relation ($\tilde{\leq}, \tilde{=}, \tilde{\geq}, \tilde{\in}$). The third is a fuzzy interval right-hand side whose semantic is gradual set belonging. Let us look at each of these one at a time.

5.2.1.1 Fuzzy Meaning of Optimize

A fuzzy meaning of "optimize" generates a fuzzy relation as follows. A fuzzy "optimize" is a target. Its meaning is "come as close as possible" to the target. There are three types of targets each of which has a translation.

1. The first is come as close as possible in the sense of equality which is then translated into a fuzzy equality relationship.

2. The second is come as close as possible or exceed in the sense of greater or equal which is translated into a fuzzy greater or equal relationship.

3. The third is come as close as possible but not surpass in the sense of less or equal which is translated into a fuzzy less or equal to relationship.

Any of these types of objective functions are moved into the fuzzy constraint set. The objective function is replaced by maximizing the target attainment. That is, a fuzzy optimize is considered as a goal. The value of the target or goal may be a subjective decision or may be a deterministic result of an optimization under deterministic constraints. Once a target or goal for the objective function has been identified, it may be translated into a flexible constraint. For example, if a deterministic objective function were

$$\min z = f(c, x) = c^T x,$$

and the target or goal were calculated or set at z_*, then this would be translated as follows.

$$f(c, x) = c^T x \leq \tilde{z}_*$$

$$\updownarrow$$

$$\max \alpha \tag{5.4}$$
$$\alpha d_0 + c^T x \leq z_* + d_0$$
$$0 \leq \alpha \leq 1,$$

where d_0 is the amount over z_* that would tolerated as a minimum. Thus, flexible objective functions may be considered as flexible constraints.

5.2.1.2 Fuzzy Relationship in the Constraints

A fuzzy relationship generates a fuzzy set belonging $\tilde{\in}$ relation. Fuzzy relationships can be equality or inequality relationships in which case we need to translate them into a fuzzy set membership function. However, the relationship may be given directly as a fuzzy set belonging in which case it does not need to be translated. Our focus here is how to translate a fuzzy relationship that is not directly given as a fuzzy membership function into a fuzzy membership function. There are two approaches.

1. The first is a fuzzy relationship that needs translating into a fuzzy membership function we call a *soft constraint* relationship, a term used in the fuzzy set theory literature. However, here we limit its meaning to that of *equality and inequality relationships with flexibility*.
2. The second type of fuzzy relationship occurs when the right-hand side is a fuzzy interval with semantics tied to *gradual set belonging*, an aspiration which is described by a fuzzy interval. Such a fuzzy interval indicates preferences as delineated by the fuzzy interval function.

For example, in a soft constraint or fuzzy interval right-hand side, one may have a fixed amount b of resources (material, time, etc.) available in a production model. However, when this value b is flexible or "soft", it means that we have an addition increment, say $b + d$ where the increment might be hiring temporary workers or purchasing additional materials from a competitor (at additional costs). The preference is to use less than or equal to b but never exceed $b + d$. In between b and $b + d$, an appropriate preference function is given, typically a linear function unless there is more information indicating the shape of the preference (membership) function. Flexibility modeled by a fuzzy interval means that the right-hand value is flexible and is explicitly given as a fuzzy interval and not as an additional increment of flexibility. Essentially, these two are the same except for one we have to construct the fuzzy interval and for the other, the fuzzy interval is already given. Computationally, we have two ways to handle flexibility. Let us look at these one at a time.

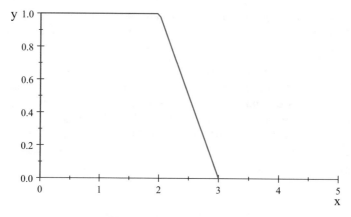

Fig. 5.1 Soft constraint $\le b = 2$ but not $\ge b + d = 3$

5.2.1.3   Soft Constraints—Implicit

A soft constraint relation is characterized by flexibility in what the relationship means. Relationships that are associated with aspirations or flexibility are fuzzy. For example, suppose we allow $\le$ to mean that we are 100% satisfied with the inequality being less than b_i but under no circumstances can we exceed $b_i + d_i$. This flexibility is a function and is depicted in Fig. 5.1. So, one considers as the acceptable values of x to belong to the interval $[0, b_i + d_i]$ but preferentially distributed in the interval. From the y-axis point of view, we want $x-$values whose corresponding $y-$values are as close to 1 as possible (x is in the interval $[0, b_i]$). For example, if $b = 2, d = 1$, then this linear soft constraint function is depicted in Fig. 5.1. That is, given $\tilde{b}$ is the trapezoidal fuzzy number $0/0/b_i/b_i + d_i$ (for example, like the one depicted in Fig. 5.1), let A_i denote the ith row, $i \in \{1, 2, \ldots, m\}$ (with $n-$components). Then,

$$A_i x \le \tilde{b}_i$$
$$\updownarrow \text{ translation}$$
$$\max \alpha$$
$$\alpha d_i + A_i \cdot x \le b_i + d_i \tag{5.5}$$
$$0 \le \alpha \le 1.$$

5.2.1.4   Fuzzy Interval—Explicit

A right-hand side which is fuzzy means that $\tilde{b}$ represents an aspiration or goal. We can interpret $\tilde{b}$ as generating a fuzzy relationship $\tilde{\le}$ (or $\tilde{=}$ in case of an equality constraint). Let $\tilde{b}$ be a one-dimensional fuzzy interval vector (n-dimensional will work in the same way component by component). There are two ways, computationally, within the flexibility semantic, to deal with this case of an explicitly given fuzzy interval representing a flexible relation.

1. The first way is to handle an explicit fuzzy interval on the right-hand side is like we did for a soft constraint (5.5) except we do not have to first construct the fuzzy interval. If the fuzzy interval is a trapezoidal number explicitly given as $0/0/b_i/b_i + d_i$, we would use it in (5.5). If the explicitly stated fuzzy interval right side is given by a trapezoidal fuzzy interval $\tilde{b} = b_i^- - d_i^- /b_i^- /b_i^+ /b_i^+ + d_i^+$ where $d_i^-, d_i^+ \geq 0$, and $b_i^- - d_i^- \leq b_i^- \leq b_i^+ \leq b_i^+ + d_i^+$ (if $b_i^- = b_i^+$ we have a triangular fuzzy number). In this more general case, we have

$$A_i x \leq \tilde{b}_i$$

$$\updownarrow \text{ translation}$$

$$\max \alpha$$

$$\alpha d_i^+ + A_i \cdot x \leq b_i^+ + d_i^+$$
$$-\alpha d_i^- + A_i \cdot x \geq b_i^- - d_i^-.$$

Other fuzzy interval forms are handled similarly.

2. The second way we will deal with a fuzzy interval right-hand side whose semantics are tied to gradual set belonging is to use a surprise function developed by Neumaier [7] and implemented in an optimization problem in [8].

Suppose we are given

$$\text{hard } y_i = (A\vec{x})_i \leq b_i \Rightarrow \text{fuzzy } y_i = A_i \cdot x \leq \tilde{b}_i, \tag{5.6}$$

where the right-hand side values of the soft constraint are fuzzy intervals and the fuzzy interval right-hand side (5.6) has membership function, $\mu_i(x)$. The associated surprise function (see [7, 8]) is,

$$s_i(x) = \left(\frac{1}{\mu_i(x)} - 1 \right)^n, n \geq 2, \tag{5.7}$$

where $n = 2$ usually suffices. The larger the n is, the "faster" the acceleration of the "edges", the values that are infeasible or close to infeasible, go toward infinity. However, the greater the n, the greater the number of computations. With $n = 2$, the function $s_i(x)$ is convex and differentiable and involves but four computations and so is amenable to efficient differential-based optimization methods. Depicted in Fig. 5.2, is a trapezoidal fuzzy number associated with preferential radiation values for a tumor cell.

A surprise function associated with a trapezoidal number like that depicted in Fig. 5.2 needs to be translated into a surprise function. This process in the context of radiation of a tumor is depicted by Fig. 5.3. A set of fuzzy interval right-hand value constraints from the surprise point of view are aggregated (intersected) into

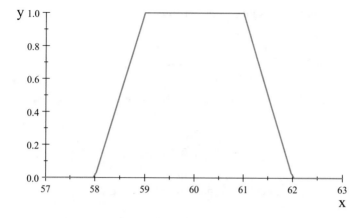

Fig. 5.2 Tumor radiation membership function

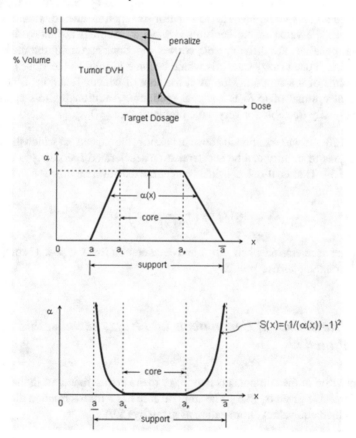

Fig. 5.3 Surprise function for tumor radiation preferrence

one function which is the sum or total surprise

$$S(x) = \sum_i s_i((A\vec{x})_i) = \sum_i \left(\frac{1}{\mu_i\,(A_i \cdot x)} - 1 \right)^2 . \tag{5.8}$$

A "best" solution based on the surprise function is given by the nonlinear optimization problem

$$\min \; z = S(x) = \sum_i \left(\frac{1}{\mu_i\,(A_i \cdot x)} - 1 \right)^2 , \tag{5.9}$$

$$\text{subject to } x \in \Omega \text{ (possibly hard constraints).} \tag{5.10}$$

The salient feature is that surprise uses a dynamic penalty for falling outside preferred membership values of one. That is, we do not have to specify an a-prior penalty for each constraint violation. The total surprise function does this automatically according to the value of the decision variable x. Another advantage is that the individual penalties are differentiable convex functions assuming the membership functions are indeed fuzzy intervals which become infinite as the values approach the endpoints of the support. Moreover, the sum of convex functions is convex so that a local optimal of (5.8) is a global optimum. Additionally, this approach is computationally tractable for very large problems (see [8]).

Remark 168 As was mentioned above in passing, the "speed" at which the surprise function goes to infinity can be accelerated or decelerated according to the power used for (5.7). That is, if one considers a generalized surprise function to be

$$s_i(x) = \left(\frac{1}{\mu_i\,(x)} - 1 \right)^p , \tag{5.11}$$

then, we get acceleration when $p > 1$ and deceleration for $0 < p < 1$, compared to $p = 1$. Our examples use $p = 2$.

5.2.2   Translation of Components into Fuzzy Membership Functions

The translation of the components into fuzzy membership functions in the case of a soft constraint is given by (5.5). The surprise case has a transformation directly into its real-valued equivalent optimization as given by (5.9).

5.2.3 Transformation of the Simultaneous Fuzzy Memberships into a Real-Valued Mathematical Programming Problem

The transformation of (5.1) and (5.2) into an optimization problem, putting all of our components together, can be symbolically denoted by the mapping $T : \mathbb{R}^{n+1} \to \mathbb{R}$ defined by:

$$\underset{x}{opt} \; T \left(\begin{matrix} f(c, x) \\ x \; \tilde{\in} \; \tilde{X} \end{matrix} \right) = \underset{x}{opt} \left[\begin{matrix} T_1(f(c, x)) \\ T_2(x \; \tilde{\in} \; \tilde{X}) \end{matrix} \right] \tag{5.12}$$

$$= \underset{x}{opt} \left[\begin{matrix} F(c, x) \; R \; b_0 \in \mathbb{R} \\ x \in \Omega \subseteq \mathbb{R}^n \end{matrix} \right] . \tag{5.13}$$

How to compute $F(c, x) \; R \; b_0$ via T_1 and Ω via T_2 is given below. However, to make this process clear, the general fuzzy optimization problem (5.1), (5.2) is restricted to the fuzzy linear programming model. Nonlinear optimization problems are done similarly.

$$\max \; z \; \tilde{\geq} \; c^T x$$
$$\text{subject to:} Ax \; \tilde{\leq} \; b \tag{5.14}$$
$$x \geq 0.$$

In this context, the *objective* $\widetilde{opt} \; z = f(c, x)$ becomes, for maximization,

$$z \; \tilde{\geq} \; c^T x. \tag{5.15}$$

The *constraint* $x \; \tilde{\in} \; \tilde{X}$ is typically

$$\tilde{X} = \{x | Ax \; \tilde{\leq} \; b\} \cap \{x \geq 0\}. \tag{5.16}$$

Example 169 (*Flexible Optimization*) Suppose we have the real-valued linear programming problem

$$\max z = x_1 + x_2 \tag{5.17}$$
$$x_1 + 4x_2 \leq 16 \tag{5.18}$$
$$x_1 + x_2 \leq 10 \tag{5.19}$$
$$x_1, x_2 \geq 0.$$

Now, suppose the flexibility in the first constraint (5.18) is 3 and in the second constraint (5.19) is 1. According to our notation, $d_1 = 3$ and $d_2 = 1$. Moreover, suppose our goal is to come as close as possible to $\hat{z} = 15$ where the optimal value of the given problem is $z^* = 10$, $x_1^* = 8$, $x_2^* = 2$. This means that $d_0 = 5$ and the objective function becomes (exceed as much as possible but never go below the optimal value of 10)

$$x_1 + x_2 \geq 15 - (1 - \alpha)5. \tag{5.20}$$

The mapping T_1 is

$$T_1(x_1 + x_2 \gtrsim 15) = 15 - (1 - \alpha)5 \tag{5.21}$$

Thus, our model translates into

$$\max \hat{z} = \alpha \tag{5.22}$$
$$-x_1 - x_2 \leq -15 + (1 - \alpha)5$$
$$x_1 + 4x_2 \leq 16 + (1 - \alpha)3$$
$$x_1 + x_2 \leq 10 + (1 - \alpha)1$$
$$0 \leq \alpha \leq 1, x_1 \geq 0, x_2 \geq 0.$$

This means that T_2 is

$$T_2 \left(\begin{bmatrix} 1 & 4 \\ 1 & 1 \end{bmatrix} \begin{bmatrix} x_1 \\ x_2 \end{bmatrix} \lesssim \begin{bmatrix} 16 \\ 10 \end{bmatrix} \right)$$

$$\Rightarrow \begin{bmatrix} 3 & 1 & 4 \\ 1 & 1 & 1 \end{bmatrix} \begin{bmatrix} \alpha \\ x_1 \\ x_2 \end{bmatrix} \leq \begin{bmatrix} 19 \\ 11 \end{bmatrix}, \tag{5.23}$$

which is the matrix representation of (5.22). In all, the translation of this problem into its real-valued equivalent, where $x_3 = \alpha$, is

$$\max \hat{z} = 0x_1 + 0x_2 + x_3$$
$$-x_1 - x_2 + 5x_3 \leq -10$$
$$x_1 + 4x_2 + 3x_3 \leq 19 \tag{5.24}$$
$$x_1 + x_2 + x_3 \leq 11$$
$$x_1, x_2, x_3 \geq 0 \text{ and } x_3 \leq 1.$$

The maximum of the flexible problem (5.24) occurs for $\alpha = \frac{1}{6} = 0.16\ldots, x_1 = \frac{149}{18} = 8.27\ldots, x_2 = \frac{23}{9} = 2.55\ldots$ with the flexible objective function value of $z = \frac{195}{18} = 10.83\ldots$ which is more than the deterministic, non-flexible, optimal value of $z^* = 10$ as expected. We have violated our original first constraint (5.18) by 2 and our second constraint (5.19) by 0.66...which is what an α different than 1 gives.

In the context of Example 169, the general transformation T_1 of (5.12), for $n = 2$, is (5.21), and the general transformation T_2 of (5.12), for $n = 2$, is (5.23).

5.3   Three Generic Flexible Optimization Model Types

We next outline the three types of flexible optimization models. The three types are:

1. Decision making in a fuzzy environment;
2. Optimization which returns fuzzy interval solutions;
3. Surprise function optimization based on transitional semantics of right-hand side fuzzy interval values.

The first category has three approaches of historical significance since they were the basis for subsequent flexible optimization types. The second category of the classification above is unique in that its solutions are fuzzy intervals. However, we will show that this approach can be considered as a subset of optimization in a fuzzy environment. The third category has been used in solving "industrial strength" problems. Very large flexible optimization models using the surprise methodology have been solved for up to 750,000 (see [9]). For a comprehensive discussion, the reader is directed to Untiedt [10] who has a review of the variety of flexible (and generalized uncertainty) optimization approaches and uses a radiation therapy example problem to compare the five different approaches which is also summarized in [11]. Moreover, Lodwick and Untiedt in [12] give a broader view of flexible and possibilistic (generalized uncertainty) optimization types. Here, we present the types according to their historical introduction in addition to two other methods. The "historical" methods are now "classical" types and broadly speaking, they represent much of what is found in the literature in one form or another. The detailed analysis of what each type does to a particular radiation therapy problem, as mentioned, is found in [10].

What follows assumes that we have obtained all the fuzzy membership functions associated with the flexible optimization problem according to what we have already presented.

5.3.1   Decision-Making in a Fuzzy Environment

The three approaches presented next, were historically, the first three approaches to solve the fuzzy optimization problem. They are given in the order in which they were published.

5.3.1.1   Bellman–Zadeh

It is assumed that we have translated our problem's flexibility into fuzzy membership functions. The objective function, Bellman and Zadeh [1] called a goal whose membership function they denoted by $\mu_G(x)$ and m fuzzy constraints to the problem they denoted $\mu_{C_i}(x)$, $i = 1, \ldots, m$. Bellman and Zadeh [1] simply intersect (via the min operator) obtaining a new membership function they denoted μ_D and then they maximize this membership function, μ_D. They called this decision making in a fuzzy environment. That is, according to their development, optimization in a fuzzy environment (one type of flexible optimization) is given by

$$\max_{x \in domain} \left\{ \mu_D(x) = \min \left\{ \mu_{C_1}(x), \mu_{C_2}(x), \dots, \mu_{C_m}(x), \mu_G(x) \right\} \right\}. \qquad (5.25)$$

For one objective function goal and one constraint given by

$$\mu_G(x) = \begin{cases} 0 \text{ if } x < 1 \text{ or } x > 3 \\ x - 1 \text{ if } 1 \le x \le 2 \\ -x + 3 \text{ if } 2 \le x \le 3 \end{cases},$$

$$\mu_C(x) = \begin{cases} 0 \text{ if } x < 2 \text{ or } x > 4 \\ x - 2 \text{ if } 2 \le x \le 3 \\ -x + 4 \text{ if } 3 \le x \le 4 \end{cases},$$

and then using the min operator on these two membership functions, we have

$$\mu_D = \min \{\mu_C(x), \mu_G(x)\}$$
$$= \begin{cases} 0 \text{ if } x < 2 \text{ or } x > 3 \\ x - 2 \text{ if } 2 \le x \le \frac{5}{2} \\ -x + 3 \text{ if } \frac{5}{2} \le x \le 3 \end{cases}.$$

The goal and constraint membership functions can be seen graphically in Fig. 5.4. The minimum of $\mu_D(x)$ and $\mu_G(x)$ is depicted in Fig. 5.5. This example has the solution

$$\max_{x \in \mathbb{R}} \mu_D(x) = \max_{x \in \mathbb{R}} \{\min \{\mu_C(x), \mu_G(x)\}\} = 0.5$$

at $x^* = 2.5$ as can be seen in the Fig. 5.5.

Example 170 Consider the flexible optimization problem given by Example 169 where we begin with (5.22).

$$\max \hat{z} = \alpha$$
$$-x_1 - x_2 \le -15 + (1 - \alpha)5$$
$$x_1 + 4x_2 \le 16 + (1 - \alpha)3$$
$$x_1 + x_2 \le 10 + (1 - \alpha)1$$
$$0 \le \alpha \le 1, x_1 \ge 0, x_2 \ge 0.$$

The solution to this problem (see Example 169) occurs for $\alpha = \frac{1}{6} = 0.16\dots$, $x_1 = \frac{149}{18} = 8.27\dots$, $x_2 = \frac{23}{9} = 2.55\dots$ with the flexible objective function value of $z =$

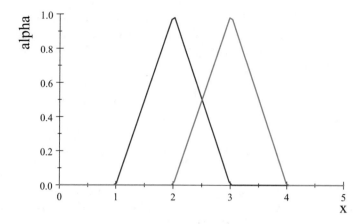

Fig. 5.4 Bellman/Zadeh fuzzy optimization $\mu_C(x)$ and $\mu_G(x)$

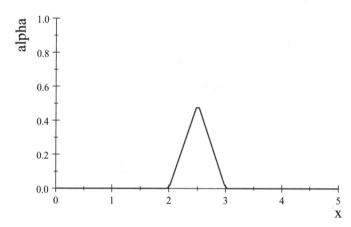

Fig. 5.5 Decision membership function $\mu_D(x)$

$\frac{195}{18} = 10.83\ldots$ which is more than the deterministic, non-flexible, optimal value of $z^* = 10$ as expected.

5.3.1.2   Tanaka, Okuda, Asai

The first actualization of the Bellman–Zadeh proposal for optimization in a fuzzy environment (5.25) in the sense of a concrete computational method was that of Tanaka, Okuda, and Asai [2, 13]. It is interesting that Tanaka, Okuda, and Asai were one of the first to correctly identify the importance in the relationship between fuzzy membership functions, their α–level, and the mathematical analysis of fuzzy entities. We note that two years after the Tanaka, Okuda, Asai publication, Negoita and

Ralescu, in their 1975 book [14], prove that a fuzzy membership function uniquely defines their corresponding α−level and a set of (nested) α−level uniquely defines the membership function, that is, a fuzzy set. In flexible optimization, Tanaka, Okuda, and Asai used an α−level approach as follows. Let

$$\mu_D(x) = \min \left\{ \mu_{c_1}(x), \mu_{c_2}(x), \ldots, \mu_{c_m}(x), \mu_G(x) \right\}, \tag{5.26}$$

then

$$\max_{x \in domain} \mu_D(x) = \sup_{\alpha \in [0,1]} \left\{ \alpha \wedge \max_{x \in C_\alpha} \mu_G(x) \right\} \tag{5.27}$$

where C_α is the α−level of the single constraint set C membership function. Recall that the α−level set $[C]_\alpha$ of $\tilde{C}$ is defined by $[C]_\alpha = \{x \in X \mid \mu_{\tilde{C}}(x) \geq \alpha\}$. This approach (5.27) is well-defined under the following conditions.

1. $\mu_G(x)$ is continuous.
2. C_α is closed and bounded $\forall \alpha \in [0, 1]$, that is, an interval.

The general problem with m−constraints, when the aggregation operator in (5.26) is the classical min operator, was proved by them to be the following.

Theorem 171 (see [13, 15] for the proof) *For m constraints,*

$$\max_{x \in domain} \mu_D(x)$$

$$= \sup_{\alpha_1, \alpha_2, \ldots, \alpha_m \in [0,1]} \left\{ \min \left(\alpha_1, \alpha_2, \ldots, \alpha_m, \max_{x \in [\tilde{C}_1]_{\alpha_1} \cap [\tilde{C}_2]_{\alpha_2} \cap \cdots \cap [\tilde{C}_m]_{\alpha_m}} \mu_{\tilde{G}}(x) \right) \right\} \tag{5.28}$$

$$= \sup_{\alpha \in [0,1]} \left\{ \min \left(\alpha, \max_{x \in [\tilde{C}_1]_\alpha \cap [\tilde{C}_2]_\alpha \cap \cdots \cap [\tilde{C}_m]_\alpha} \mu_{\tilde{G}}(x) \right) \right\} \tag{5.29}$$

where $[\tilde{C}_k]_{\alpha_k} = C_k(\alpha_k) = \left\{ x \mid \mu_{C_k}(x) \geq \alpha_k \right\}$ the α_k−level set of the kth constraint.

Remark 172 To insure that the problem is well-defined, we need the following to hold. (1) μ_G has to be continuous. (2) C_α has to be closed and bounded. For μ_G to be continuous, which, in the form $f(c, x)$ of equation (5.1), f has to be continuous in the parameters c and variable x and $g(a, x)$ of (5.3) is also continuous in the parameters a and the variable x where it is assumed that a and c are fuzzy intervals. This means that all alpha levels of the parameters (a and c) are closed and bounded intervals. Thus, $f(c, x)$ and $g(a, x)$ are fuzzy intervals for x in a compact domain.

Definition 173 (*see* [13]) Let

$$\lim_{\alpha_n \to \alpha} C_{\alpha_n} = C_\alpha.$$

A function f is α−**continuous** if

$$\lim_{\alpha_n \to \alpha} \left\{ \max_{x \in C_{\alpha_n}} f(x) \right\} = \max_{x \in C_\alpha} f(x).$$

For C_α a closed and bounded interval, then there exists $x^* \in C_\alpha$ such that

$$f(x^*) = \max_{x \in C_\alpha} f(x).$$

From the Tanaka, Okuda, and Asai point of view, the flexible optimization problem is to find α^* and x^* such that

$$\alpha^* \wedge f(x^*) = \sup_{\alpha \in [0,1]} \left\{ \alpha \wedge \left(\max_{x \in C_\alpha} f(x) \right) \right\}.$$

Now α^* is a unique optimal if and only if

$$\alpha^* = \max_{x \in C_{\alpha^*}} f(x). \tag{5.30}$$

Thus the question is what are the conditions for which (5.30) holds. They are as follows.

Theorem 174 *If α^* is unique and f is α-continuous, then α^* is optimal if and only if*

$$\alpha^* \in \max_{x \in C_{\alpha^*}} f(x)$$

which means

$$\sup_\alpha \left\{ \alpha \wedge \left(\max_{x \in C_\alpha} f(x) \right) \right\} = \max_{x \in T} f(x)$$

where

$$T = \{ x | h(x) = f(x) - \mu_C(x) = f(x) - g(x) = 0 \}.$$

Remark 175 The conditions needed to have α^* unique is the strictly concavity and α-continuity of f. Given these conditions for f, the flexible optimization problem becomes a problem of finding the roots of $h(x)$.

These conditions result in an algorithm that Tanaka, Okuda, and Asai developed.

Algorithm 176 *Step 0: Choose $k = 1$, a tolerance $\epsilon > 0$, any $\alpha_1 \in (0, 1]$, and step size $r \in (0, 1)$.*
Step 1: Compute

$$C_k = C_{\alpha_k}.$$
$$f_k = \max_{x \in C_k} f(x)$$
$$\epsilon_k = \alpha_k - f_k$$

Step 2: If

$$\epsilon_k \leq \epsilon, \ go \ to \ END.$$

Step 3: Set

$$\alpha_{k+1} = \alpha_k - r\epsilon_k, \ go \ to \ Step \ 1.$$

END: $\alpha^* = \alpha_k$, x^* *is obtained by solving*

$$\max_{x \in C_{\alpha^*}} f(x).$$

Remark 177 When we choose $r = \frac{1}{2}$, we have bisection. Of course, finding α^* is a root finding or fixed point problem. Therefore, a "speed up" of the algorithm is possible by using a preconditioned Newton's Method.

5.3.1.3 Zimmermann

Arguably, the most used flexible optimization approach was developed by Zimmermann [16]. This was historically the second implementation of the Bellman–Zadeh decision making in a fuzzy environment. As can be seen, it is an intuitive and straight forward approach. It is, on one level, simpler than the Tanaka, Okuda, and Asai implementation. Essentially, we have already presented Zimmermann's approach and illustrate it as follows.

Example 178 Consider the flexible optimization problem given by Example 169 which is in fact optimized according to Zimmermann's original approach. That is,

$$\max \hat{z} = \alpha$$
$$-x_1 - x_2 \leq -15 + (1 - \alpha)5$$
$$x_1 + 4x_2 \leq 16 + (1 - \alpha)3$$
$$x_1 + x_2 \leq 10 + (1 - \alpha)1$$
$$0 \leq \alpha \leq 1, x_1 \geq 0, x_2 \geq 0.$$

is the translation of the flexible optimization with $d_0 = 5$, $d_1 = 3$, $d_2 = 1$. The result of the optimization problem above is, as we have already seen,

$$z^* = 10.83\ldots = \left(\frac{149}{18}\right) + \left(\frac{23}{9}\right) = \frac{195}{18}$$

on the original objective of $z = x_1 + x_2$, where $\alpha = \frac{1}{6} = 0.16\ldots$, $x_1 = \frac{149}{18} = 8.27\ldots$, $x_2 = \frac{23}{9} = 2.5\ldots$.

Zimmermann requires that every constraint be related to a single alpha level. That is, all alpha levels of each of the membership functions are at the same level. Clearly, this is not a Pareto optimal.

Example 179 Allow each of the constraints to have their own alpha level and value violations of each of our constraints differently, the optimization becomes

$$
\begin{bmatrix}
z = \max c_1 \alpha_1 + c_2 \alpha_2 + c_3 \alpha_3 \\
\text{subject to: } -x_1 - x_2 + 5\alpha_1 \leq -10 \\
x_1 + 4x_2 + 3\alpha_2 \leq 19 \\
x_1 + x_2 + \alpha_3 \leq 11 \\
x_1, x_2, \alpha_1, \alpha_2, \alpha_3 \geq 0, \alpha_1, \alpha_2, \alpha_3 \leq 1
\end{bmatrix}
$$

where c_1, c_2, c_3 are cost coefficients which we can vary. When $c_1 = c_2 \gg c_3$ we can obtain the solution of

$$
z^* = 11 = \left(\frac{28}{3}\right) + \left(\frac{5}{3}\right) = \frac{33}{3}
$$

where $\alpha_1^* = \frac{8}{15} = 0.53\ldots, \alpha_2^* = 1, \alpha_3 = 0$ and $x_1^* = \frac{28}{3} = 9.3\ldots, x_2^* = \frac{5}{3} = 1.6\ldots$. This example illustrates that the Zimmermann approach is not optimal according to the Pareto criterion.

5.3.2 Flexible Optimization that Returns Fuzzy Interval Solutions

The second approach in our classification returns solutions that are fuzzy intervals rather than real-number solutions. The solutions examined so far for the fuzzy programming problem have been real-valued solutions. Ralescu [17] first suggested that a fuzzy problem should have a fuzzy solution, and Verdegay [4] proposes a method for obtaining a fuzzy solution. However, in fuzzy interval optimization, it was Verdegay who first considers a problem with fuzzy constraints,

$$
z = f(x) \tag{5.31}
$$
$$
x \subseteq \tilde{C}, \tag{5.32}
$$

where the set of constraints have a membership function μ_C, with alpha-cuts $\tilde{C}_\alpha$. The containment form of the constraint set as given by (5.32) was originally suggested, in the context of interval optimization, by [18, 19] and then again by [20] in the context of fuzzy optimization.

Verdegay defines x_α as the set of solutions that satisfy constraints $\tilde{C}_\alpha$. Then a fuzzy solution to the flexible optimization problem is

$$\max_{x \in \tilde{C}_\alpha} z_\alpha = f(x) \tag{5.33}$$

$$0 \leq \alpha \leq 1.$$

Verdegay proposes solving (5.33) parametrically for $\alpha \in [0, 1]$ to obtain a fuzzy solution $\tilde{x}$, with α-cut x_α, which yields fuzzy objective value $\tilde{z}$, with α-cut z_α.

It should be noted that the (crisp) solution obtained via Zimmermann's method corresponds to Verdegay's solution in the following way. If the objective function is transformed into a goal, with membership function μ_G, then Zimmermann's optimal solution, x^*, is equal to Verdegay's optimal value $x(\alpha)$ for the value of α which satisfies

$$\mu_G(z_\alpha^* = cx_\alpha^*) = \alpha.$$

In other words, when a particular α-cut of the fuzzy solution, (x_α) yields an objective value $(z_\alpha = cx_\alpha)$ whose membership level for goal G, $(\mu_G(z_\alpha))$ is equal to *the same* α, then that solution x_α corresponds to Zimmermann's optimal solution, x^*.

5.3.3 Fuzzy Interval Right-Hand Sides

The third approach of our classification uses surprise functions. This is a newer approach that is useful in large "industrial-sized" optimization problems. The use of surprise functions for right sides that are fuzzy intervals whose meaning is tied to gradual set belonging. Once the fuzzy interval membership function has been determined, it is transformed into a surprise function. The sum of the surprise functions are then minimized. It is should be clear from the example how to proceed in general.

Example 180 Consider the flexible optimization problem given by Example 169 but restricted to the first and second constraints being fuzzy intervals. In this context we have

$$\max z = x_1 + x_2$$
$$\text{Subject to:}$$
$$x_1 + 4x_2 \leq \mu_{16}(x_1, x_2)$$
$$x_1 + x_2 \leq \mu_{10}(x_1, x_2)$$
$$x_1, x_2 \geq 0,$$

where we have for the first constraint, the membership function is given by

$$\mu_{16}(x_1, x_2) = \begin{cases} 1 & \text{if } x_1 + 4x_2 \leq 16 \\ -\frac{1}{3}(x_1 + 4x_2) + \frac{19}{3} & \text{if } 16 \leq x_1 + 4x_2 \leq 19 \\ 0 & \text{if } x_1 + 4x_2 \geq 19 \end{cases},$$

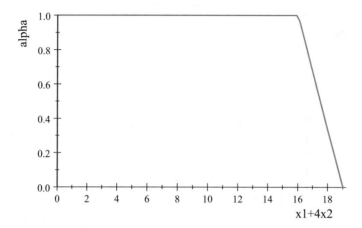

Fig. 5.6 $\mu_{16}(x_1, x_2)$

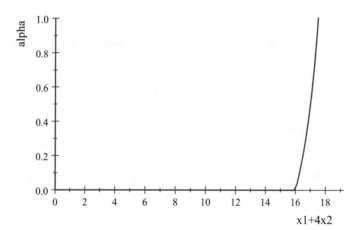

Fig. 5.7 Surprise for $\mu_{16}(x_1, x_2)$

and whose trapezoidal graph is Fig. 5.6. The corresponding surprise function is

$$s_{16}(x_1, x_2) = \left(\frac{1}{\mu_{16}(x_1, x_2)} - 1 \right)^2 = \begin{cases} 0 \text{ for } 0 \leq x_1 + 4x_2 \leq 16 \\ \left(\frac{1}{-\frac{1}{3}(x_1+4x_2)+\frac{19}{3}} - 1 \right)^2 \text{ for } 16 \leq x_1 + 4x_2 \leq 19 \end{cases}$$

whose graph is Fig. 5.7. For the second constraint, the membership function is given by

$$\mu_{10}(x_1, x_2) = \begin{cases} 1 \text{ if } x_1 + x_2 \leq 10 \\ -x_1 - x_2 + 11 \text{ if } 10 \leq x_1 + x_2 \leq 11 \\ 0 \text{ if } x_1 + x_2 \geq 11 \end{cases},$$

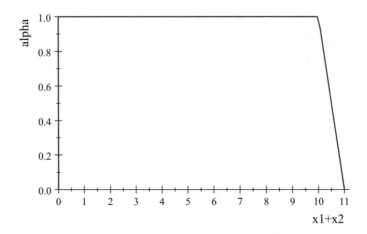

Fig. 5.8 $\mu_{10}(x_1, x_2)$

and whose trapezoidal graph is Fig. 5.8 The corresponding surprise function is

$$s_{10}(x_1, x_2) = \left(\frac{1}{\mu_{10}(x_1, x_2)} - 1 \right)^2 = \begin{cases} 0 \text{ for } 0 \leq x_1 + x_2 \leq 10 \\ \left(\frac{1}{-(x_1+x_2)+11} - 1 \right)^2 \text{ for } 10 \leq x_1 + x_2 \leq 11 \end{cases}$$

whose graph is Fig. 5.9. The problem becomes

$$\max z = x_1 + x_2 - \left(\frac{1}{\mu_{16}(x_1, x_2)} - 1 \right)^2 + \left(\frac{1}{\mu_{10}(x_1, x_2)} - 1 \right)^2$$

Subject to: (5.34)

$$x_1 + 4x_2 \leq 19$$
$$x_1 + x_2 \leq 11$$
$$x_1, x_2 \geq 0$$

which is a nonlinear programming problem (where we have changed the sign of the original objective function surprise part).

5.4   Summary

This section developed the components of flexible optimization clearly linking the semantics and the construction of the fuzzy constraint set with the fuzzy objective function via the interpretation of what a fuzzy relationship is. The translation of flexibility into optimization models was given.

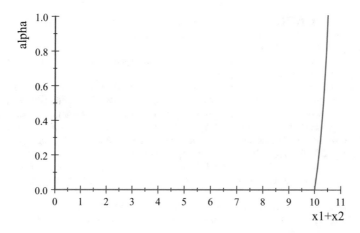

Fig. 5.9 Surprise for $\mu_{10}(x_1, x_2)$

5.5 Exercises

Exercises 181 Consider the flexible optimization problem given by Example 169. Solve this problem using the Tanaka-Asai method, with $\alpha_1 = \frac{1}{2}$, $\epsilon = 0.01$, and $r = \frac{1}{2}$.

Exercises 182 Consider the flexible optimization problem given by Example 169. Solve this problem using the Verdegay method (5.33).

Exercises 183 Complete Example 180. In particular, optimize problem (180).

Exercises 184 Prove the theorems whose proofs were not given.

References

1. R.E. Bellman, L.A. Zadeh, Decision-making in a fuzzy environment. Manag. Sci. Ser. B **17**, 141–164 (1970)
2. H. Tanaka, T. Okuda, K. Asai, On fuzzy mathematical programming (1973)
3. H. Zimmermann, Optimization in fuzzy environment, in *Paper presented at the XXI TIMS and 46th ORSA Conference* (Puerto Rico Meeting, San Juan, Puerto Rico, 1974)
4. J.L. Verdegay, Fuzzy mathematical programming, in *Fuzzy Information and Decision Processes*, ed. by M.M. Gupta, E. Sanchez (North Holland Company, Amsterdam, 1982), pp. 231–237
5. M. Delgado, J. Kacprzyk, J.-L. Verdegay, M.A. Vila, *Fuzzy Optimization: Recent Advances* (Physica-Verlag, Heidelberg, 1994)
6. M. Delgado, J.-L. Verdegay, M.A. Vila, A general model for fuzzy linear programming. Fuzzy Sets Syst. **29**, 21–29 (1989)
7. A. Neumaier, Fuzzy modeling in terms of surprise. Fuzzy Sets Syst. **135** (2003)

8. W.A. Lodwick, A. Neumaier, F.D. Newman, Optimization under uncertainty: methods and applications in radiation therapy, in *Proceedings 10th IEEE International Conference on Fuzzy Systems 2001*, vol. 3, pp. 1219–1222

9. W.A. Lodwick, K. Bachman, Solving large scale fuzzy possibilistic optimization problems. Fuzzy Opt. Decis. Making **4**(4), 257–278 (2005)

10. E. Untiedt, Fuzzy and possibilistic programming techniques in the radiation therapy problem: an implementation-bases analysis. Masters Thesis, University of Colorado Denver (2006)

11. E. Untiedt, W.A. Lodwick, On selecting an algorithm for fuzzy optimization, in *Foundations of Fuzzy Logic and Soft Computing: 12th International Fuzzy System Association World Congress, IFSA 2007, Cancun, Mexico, June 2007, Proceedings*, eds. by P. Melin, O. Castillo, L.T. Aguilar, J. Kacpzryk, W. Pedrycz (Springer, 2008), pp. 371–380

12. W.A. Lodwick, E. Untiedt, Chapter 1: Fuzzy optimization, in *Fuzzy Optimization: Recent Developments and Applications*, ed. by W.A. Lodwick, J. Kacprzyk (Springer, New York, 2010)

13. H. Tanaka, T. Okuda, K. Asai, On fuzzy mathematical programming (1974)

14. C.V. Negoita, D.A. Ralescu, *Applications of Fuzzy Sets to Systems Analysis* (Birkhauser, Boston, 1975)

15. M. Inuiguchi, W.A. Lodwick, Foundational contributions of K. Asai and H. Tanaka to fuzzy optimization. Fuzzy Sets Syst. **274**, 24–46 (2015)

16. H. Zimmermann, Description and optimization of fuzzy systems. Int. J. Gen. Syst. **2**, 209–215 (1976)

17. D. Ralescu, Inexact solutions for large-scale control problems, in *Proceedings of the First International Congress on Mathematics at the Service of Man* (1977)

18. A.L. Soyster, Convex programming with set-inclusive constraints and applications to inexact linear programming. Oper. Res. **21**, 1154–1157 (1973)

19. A.L. Soyster, A duality theory for convex programming with set-inclusive constraints. Oper. Res. **22**, 892–898 (1974)

20. C.V. Negoita, N. Sularia, On fuzzy mathematical programming and tolerances in planning. Econ. Comput. Econ. Cybern. Stud. Res. **1**, 3–15 (1976)

Chapter 6
Generalized Uncertainty Optimization

6.1 Introduction to Generalized Uncertainty Optimization

It is clear that when the coefficients in the body of the problem and/or the right-hand side and/or the costs of an optimization problem are fuzzy intervals whose semantics are tied to information deficiency, a generalized uncertainty optimization problem ensues. Since this chapter focuses on lack of information, we are *not* dealing with gradual set belonging. Also, recall that for generalized uncertainty optimization, a decision x generates a distribution (for flexible optimization a decision x generates a real number). The reader may consult our previous chapters. In particular, the reader might look at Chap. 4, Sect. 4.4 and Examples 152 and 153.

The mathematical statement of the generalized uncertainty optimization model is

$$opt\ z = f(\hat{c}, x),\ f : \hat{H} \times \mathbb{R} \to \hat{H} \tag{6.1}$$

$$\text{subject to: } g(\hat{a}, x) \le \hat{b}, g : \hat{H} \times \mathbb{R} \to \hat{H} \tag{6.2}$$

where $\hat{H}$ is the function space of generalized functions. As stated before, the generalized uncertainty problem (6.1), (6.2) is ill-defined unless an order for the relationship $\le$ and the operation "optimize" is specified for generalized uncertainty functions. Two broad approaches have emerged to obtain an equivalent problem translated into a space where an *order* for distributions does exist. They are:

1. Fuzzy Banach space approaches such as [9–12];
2. Real Euclidean space approaches such as Tanaka, Asai [1, 2], Buckley [3, 4], Inuiguchi, Sakawa, Kume [13–17], and Lodwick, Jamison, Thipwiwatpotjana [5–8].

Each of these two broad approaches has its own way of dealing with how the associated generalized uncertainty distributions are mapped onto an ordered field which is necessary for the determination of the constraint inequalities and what is " best".

© Springer Nature Switzerland AG 2021
W. A. Lodwick and L. L. Salles-Neto, *Flexible and Generalized
Uncertainty Optimization*, Studies in Computational Intelligence 696,
https://doi.org/10.1007/978-3-030-61180-4_6

Thus, once the order relationships have been resolved, all relations ($opt, =$, $\leq$, $\in$) are those associated with real numbers, that is, our usual ones. There are three places in a generalized uncertainty optimization problem (6.1), (6.2) where uncertainty occurs.

1. The objective function rim parameters $\hat{c}$;
2. Body coefficient parameters $\hat{a}$;
3. The right-hand side rim parameters $\hat{b}$.

6.1.1   Distributions

We next review distributions since distributions characterize generalized uncertainties. We focus on two types of generalized uncertainty distributions, (1) those arising from fuzzy intervals and (2) those arising from pairs of bounding functions. The first of these are fuzzy intervals whose semantic is tied to *information deficiency*. The way to construct these has been developed in previous chapters.

6.1.1.1   Fuzzy Intervals

A fuzzy interval tied to lack, partial and/or deficient information is a possibility distribution such as that depicted in Fig. 6.1. Given a fuzzy interval, one can generate:

1. A possibility which is the fuzzy interval itself (the trapezoid 1/2/4/6 depicted in Fig. 6.1);
2. An upper possibility from a fuzzy interval (the "trapezoid" 1/2/∞/∞ depicted in Fig. 6.1) representing an optimistic view of the values of a fuzzy interval;
3. A lower necessity from a fuzzy interval (the "trapezoid" 4/6/∞/∞ depicted in Fig. 6.1) representing a pessimistic or conservative risk averse view of the values of a fuzzy intervals;
4. Any distribution between the upper and lower distribution such as the "trapezoid" 1/6/∞/∞ depicted in Fig. 6.1, the middle distribution;
5. Both upper (possibility) and lower (necessity) distributions taken together that generates a minimax optimization problem.

6.1.1.2   Pairs of Bounding Functions

The previous chapters discussed pairs of bounding functions which are either provided by the problem itself or the problem specifies enough data to construct pair of bounding functions. For example, recall that interval-valued probabilities arise when a cumulative distribution function $F(x)$ of a probability density function $p(x)$ is known to exist inside a bounding pair of functions

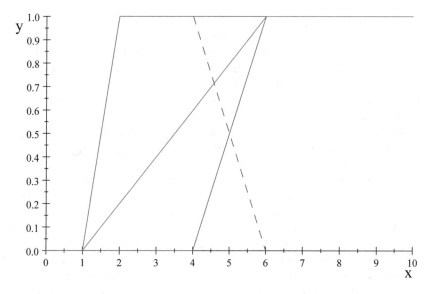

Fig. 6.1 Fuzzy interval as an uncertainty

$$F(x) \in [\underline{F}(x), \overline{F}(x)] \qquad (6.3)$$

in which case the bounding functions are the possibility/necessity pair. Bounding pairs of functions can also be constructed (see [5]) in cases in which complete information is missing to enable a single probability distribution to be obtained. These bounding functions, as we have seen, have the form

$$\Pr(X \in A) \in [Nec(A), Pos(A)] \qquad (6.4)$$

for all measurable sets A, where Pr is the probability that random variable $X \in A$. Note that (6.4) has the same structure as the necessity/possibility pairs of Fig. 6.1.

6.1.2   Fuzzy Optimization Mapped to Fuzzy Banach Spaces

The fuzzy Banach space approaches deal directly with function spaces. The solution methods compute a sequence of approximate solution functions whose limit is the solution function to which the method converges. Therefore the researchers work in a Banach space since the sequence of functions generated by their methods is convergent once the sequence of functions are shown to be a Cauchy sequence. Partial orders on generalized functions, such as integral norms or stochastic dominance approaches (see [18, 19]) need to be used. That is, fuzzy Banach space methods work in the function spaces directly.

The solution(s) to optimization problems many times involve iteration and/or approximation such as steepest descent in function spaces, the approximations are functions. It is clear that for generalized uncertainty optimization problems in function spaces, convergence analysis associated with any iterative or approximation algorithmic method is facilitated by having the iterates in a Banach space. When one keeps the iterates and/or approximations as generalized uncertainty entities in a fuzzy Banach space, each successive iterate and/or approximate is a generalized uncertainty entity (a distribution, a function) which in turn means that the convergent entity or approximation is back in the space—it is again a generalized uncertainty entity. Thus, the convergence analysis is one associated with a space of generalized uncertainty functions and once one has convergence, the result is a function and at this point a decision is made based on this distribution, this function. In the case where the problem is solved in a fuzzy Banach space, the general fuzzy optimization problem (6.1), (6.2) returns a distribution, a function. There is no translation necessary except that of inclusion into an appropriate Banach space. Diamond and Kloeden [9, 10] use a space in which their set of fuzzy sets, which they denoted E^n, over $\mathbb{R}^n$, is endowed with a neighborhood systems and a metric that renders it a Banach space. After this the Kurush–Kuhn–Tucker (KKT) conditions are worked out (see [10, 20]). Saito and Ishii [12] also develop the KKT conditions for optimization over fuzzy numbers. Diamond [21] and Jamison [11, 22] consider equivalence classes in developing a Banach space of fuzzy sets. Jamison [11] goes on to use the Banach space developed from the equivalence classes to solve optimization problems. Once the embedding to the generalized uncertainty optimization problem into the fuzzy Banach space is made, decisions are based on the resulting function solution which means that there still remains a mapping from a function to the decision space. This process is often called *defuzzification*. Prior to that, the entities that are used in the analysis come directly from the model (6.1), (6.2) as is. This approach endows the set of fuzzy sets with a metric and neighborhood system rendering it a Banach space in which optimization problems are then solved. We leave aside the Banach space approach since it involves the transformation of the space of fuzzy numbers into a Banach space which will, in general, lose an isomorphic copy of fuzzy intervals (see [23, 24]). The method developed here is based on real Euclidean spaces.

6.1.3   Real Euclidean Space Approaches

The real Euclidean space approaches are distinguished by the way they map each of the three structural types. Recall that the three structural types in $\mathbb{R}^n$ are: (1) Objective function rim coefficients, $\hat{c}$; (2) Body coefficients, $\hat{a}$; and (3) Right-hand side rim parameters, $\hat{b}$, which is where generalized uncertainty occurs in a generalized uncertainty optimization problem. That is, the real Euclidean space methods need to first map the problem from the function space $\hat{H}$ to $\mathbb{R}^n$ and the way different researchers map the problem distinguishes the various types of methods. The different Euclidean types are subsequently developed more fully. However, generally speaking, the mappings of the three structural places are the following.

1. The objective function of generalized uncertainty optimization uses a scalarized function (see [25, 26]) that maps the objective function of generalized distributions into real numbers in a similar way that an expected value maps a probability function into real numbers. Various scalarized functions have been used. This is the approach of [7].

2. When we have body coefficients, we have four types of approaches.

   (a) The first is a fuzzy alpha-level approach of Negoita, Sularia [27], and Lodwick [28] based on the interval analysis constraint interpretation of the 1973 Soyster method [29].
   (b) The second is Tanaka, Okuda, Asai [2], who use degrees of satisfaction.
   (c) The third approach use modalities of Dubois [30], and Inuiguchi, Ichihashi, Kume [31], which are generalizations of degrees of satisfaction and work like a chance constraint of Charnes, Cooper [32] on generalized uncertainty fuzzy interval relations.
   (d) The fourth method is that of Lodwick, Jamison [7], who consider violations of constraints in a penalized way putting these violations into the objective function with the aim of minimizing these violations.

3. When the right-hand side rim parameters are a generalized uncertainty type, it is moved to the left side and considered as a body coefficient and handled by one of the methods listed above.

We will focus on three types of methods since these are distinct variations. The first set is that of Tanaka, Ichihashi, and Asai (2.b above) who look at the generalized uncertainty problem from a chance constraint point of view. The second is that of Lodwick and Jamison who take violations of constraints into the objective as penalties. And the third type is a minimax regret approach of Thipwiwatpotjana and Lodwick. An extension of minimax regret is the new (original and presented here) Luiz Salles Neto generalized multi-objective minimax regret.

6.1.3.1 An Historically Important Generalized Uncertainty Optimization Method—Tanaka, Asai, Ichihashi

Much of the material in this section can also be found in Inuiguchi, Lodwick [33]. Recall that fuzzy imprecision arises when elements of a set (for instance, a feasible set) are members of the set to varying degrees, which are defined by the membership function of the set. Possibilistic imprecision arises when elements of a set (say, a feasible set) are known to exist as either full members or non-members, but whether they are members or non-members is known with a varying degree of certainty, which is defined by the possibility distribution of the set. This possibilistic uncertainty arises from a lack of information. In the present section, we examine this historically important possibilistic programming formulation.

Tanaka and Asai [1] and Tanaka, Ichihashi, and Asai [2] in the mid-1980s, proposed a technique for dealing with ambiguous coefficients and right hand sides based upon a possibilistic definition of "greater than zero." The reader will note that this approach bears many similarities to the flexible programming proposed by Tanaka, Okuda, and Asai a decade earlier, which is discussed in Chap. 5. Indeed, the 1984 publications refer to *fuzzy* variables. This approach has subsequently been classified by Inuiguchi [13], and Lai, Hwang [34] as possibilistic programming because the imprecision it represents stems from a lack of information about the values of the coefficients.

Consider a programming problem with non-interactive possibilistic $\hat{A}$, $\hat{b}$, and $\hat{c}$, whose possible values are defined by fuzzy sets as was developed by Zadeh's initial possibilistic distribution formulation:

$$opt\, z = \hat{c}^T x \qquad (6.5)$$
$$\hat{A}x \leq \hat{b}$$
$$x \geq 0.$$

Tanaka, Ichihashi, and Asai [2] transform the problem in several steps. First, the objective function is viewed as a goal, which is the idea of satisficing found in Simon [35, 36] discussed in Chap. 1. As in flexible programming, the goal becomes a constraint with the aspiration level for the objective function on the right-hand-side of the inequality. Next, a new decision variable x_0 is added. Finally, the right-hand side rim values, the b's, are moved to the left hand side, so that the form of the possibilistic linear programming problem is

$$\hat{a}'_i x'_i \geq 0 \qquad (6.6)$$
$$i = 0, 1, 2, \ldots, m,$$
$$x' \geq 0$$

where $x' = (1, x^T)^T = (1, x_1, x_2, \ldots, x_n)^T$, and $\hat{a}_i = (\hat{b}_i, \hat{a}_{i1}, \ldots \hat{a}_{in})$.

Note that all the parameters, A, b, and c are now grouped together in the new constraint matrix $\hat{A}$. Because the objective function(s) has(have) become goals, the cost coefficients, c, are part of the constraint coefficient matrix A. Furthermore, because the right-hand side values have been moved to the left-hand side, the b's are also part of the constraint coefficient matrix. The right-hand-sides corresponding the former objective functions are the aspiration levels of the goals. So each constraint becomes

$$\hat{Y}_i = \hat{a}_i x \geq 0,$$

where

$$\hat{a}_i = (\hat{b}_i, \hat{a}_{i1}, \ldots \hat{a}_{in}).$$

" $\hat{Y}_i$ is almost positive", denoted by $\hat{Y}_i \overset{\sim}{\geq} 0$, is defined by

$$\hat{Y}_i \hat{\geq} 0 \tag{6.7}$$

$$\Updownarrow$$

$$\max h$$

$$\text{subject to: } \mu_{\hat{Y}_i}(0) \leq 1 - h, \tag{6.8}$$

$$x^t \alpha_i \geq 0.$$

The measure of the non-negativity of $\hat{Y}_i$ is h. The greater the value of h, the stronger the meaning of "almost positive". Actually, h is $1 - \alpha$, where α is the level of membership used by Bellman and Zadeh. For the Tanaka, Ichihashi, Asai approach, the problem is to maximize h.

6.1.3.2 Interval-Valued Probability Optimization

It is assumed that the data of the coefficients and/or parameters have been obtained or constructed as interval-valued probabilities

$$F(x) \in [\underline{F}(x), \overline{F}(x)],$$

which could also be any set of bounding functions, for example,

$$[pos_2(x), pos_1(x)],$$
$$[nec(x), pos(x)],$$
$$[bel(x), pl(x)],$$

and so on. Here we assume we have an IVPM. Furthermore, we will assume that we choose a function from the family of functions defined by bounding functions or use the pair of bounding functions together. Parts of this section can also be found in [7].

1. Interval-Valued Integration, Extension and Independence Three key concepts are needed for the application of Interval-Valued Probability Measures (IVPMs) to mathematical programing problems. They are integration, extension and independence which are defined next.

Definition 185 Given an F-probability field $\mathcal{R} = (S, \mathcal{A}, i_m)$ and an integrable function $f : S \to R$ we define:

$$\int_A f(x) \, di_m = \left[\inf_{p \in \mathcal{M}(\mathcal{R})} \int_A f(x) \, dp, \ \sup_{p \in \mathcal{M}(\mathcal{R})} \int_A f(x) \, dp \right] \tag{6.9}$$

It is easy to see that if f is an $\mathcal{A}$−measurable simple function such that $f(x) = \begin{cases} y & x \in A \\ 0 & x \notin A \end{cases}$ with $A \in \mathcal{A}$, then

$$\int_A f(x)\,di_m = y(i_m(A)) \tag{6.10}$$

Further, if f is a simple function taking values $\{y_k \mid k \in K\}$ on an at most countable collection of disjoint measurable sets $\{A_k \mid k \in K\}$ that is, $f(x) = \left\{ \begin{matrix} y_k & x \in A_k \\ 0 & x \notin A \end{matrix} \right|$ where $A = \cup_{k \in K} A_k$, then

$$\int_A f(x)\,di_m = \left[a^- \left(\int_A f(x)\,di_m \right), a^+ \left(\int_A f(x)\,di_m \right) \right], \tag{6.11}$$

where

$$a^+ \left(\int_A f(x)\,di_m \right) = \sup \{ \Sigma_{k \in K} y_k \Pr(A_k) \mid \Pr \in \mathcal{M}(\mathcal{R}) \} \tag{6.12}$$

and

$$a^- \left(\int_A f(x)\,di_m \right) = \inf \{ \Sigma_{k \in K} y_k \Pr(A_k) \mid \Pr \in \mathcal{M}(\mathcal{R}) \}. \tag{6.13}$$

Note that these can be evaluated by solving two linear programing problems since $\Pr \in \mathcal{M}(\mathcal{R})$ implies that $\Sigma_{k \in K} \Pr(A) = 1$ and $\Pr(\cup_{l \in L \subset K} A_l) \in i_m(\cup_{l \in L \subset K} A_l)$. In general, if f is an integrable function and $\{f_k\}$ is a sequence of simple functions converging uniformly to f, then we can determine the integral with respect to f by noting that

$$\int_A f(x)\,di_m = \lim_{k \to \infty} \int_A f_k(x)\,di_m,$$

where

$$\lim_{k \to \infty} \int_A f_k(x)\,di_m = \left[\lim_{k \to \infty} a^- \left(\int_A f_k(x)\,di_m \right), \right.$$
$$\left. \lim_{k \to \infty} a^+ \left(\int_A f_k(x)\,di_m \right) \right]$$

provided the limits exist.

Example 186 Consider the IVPM constructed from the interval $[a, b]$. Then $\int_R x\,di_m = [a, b]$, that is, the interval-valued expected value is the interval itself.

We next define an extension principle in the context of IVPMs. This extension principle generalizes the united extension of Strother [37–39], Moore, Strother, Yang [40], and Moore [41] as well as the extension principle of Zadeh [42].

Definition 187 Let $\mathcal{R} = (S, \mathcal{A}, i_m)$ be an F-probability field and $f : S \to T$ a measurable function from measurable space $(S, \mathcal{A})$ to measurable space $(T, \mathcal{B})$. Then the F-probability $(T, \mathcal{B}, l_m)$ defined by

$$l_m(B) = [\inf\{\Pr\left(f^{-1}(B)\right) \mid \Pr \in \mathcal{M}(\mathcal{R})\}, \sup\{\Pr\left(f^{-1}(B)\right) \mid \Pr \in \mathcal{M}(\mathcal{R})\}]$$

is called the **extension** of the R-probability field to $(T, \mathcal{B})$.

That this defines an F-*probability* field is clear from our earlier chapters. In addition, it's easy to see that this definition is equivalent to setting

$$l_m (A) = i_m \left(f^{-1} (A) \right)$$

which allows us to evaluate. The combination of IVPMs when the variables are independent is addressed next. The situation, when dependencies may be involved, is not discussed here.

Given measurable spaces $(S, \mathcal{A})$ and $(T, \mathcal{B})$ and the product space

$$(S \times T, \mathcal{A} \times \mathcal{B}),$$

assume $i_{X \times Y}$ is an IVPM on $\mathcal{A} \times \mathcal{B}$. Call i_X and i_Y defined by

$$i_X (A) = i_{X \times Y} (A \times T)$$

and

$$i_Y (B) = i_{X \times Y} (S \times B)$$

the marginals of $i_{X \times Y}$. The marginals, i_X and i_Y, are IVPMs.

Definition 188 Call the marginal IVPMs **independent** if and only if $i_{X \times Y} (A x B) = i_X (A) i_Y (B) \ \forall A, B \subseteq S$.

Definition 189 Let $\mathcal{R} = (S, \mathcal{A}, i_X)$ and $\mathcal{Q} = (T, \mathcal{B}, l_Y)$ be F-probability fields representing uncertain random variables X and Y. Define the F-probability field $(S \times T, \mathcal{A} \times \mathcal{B}, i_{X \times Y})$ by

$$i^+_{X \times Y} (A x B) = \sup\{\Pr_X (A) \Pr_Y (B) \mid \Pr_X \in \mathcal{M} (\mathcal{R}) , \Pr_Y \in \mathcal{M} (\mathcal{Q})\},$$

$$i^-_{X \times Y} (A \times B) = \inf\{\Pr_X (A) \Pr_Y (B) \mid \Pr_X \in \mathcal{M} (\mathcal{R}) , \Pr_Y \in \mathcal{M} (\mathcal{Q})\},$$

where $(S \times T, \mathcal{A} \times \mathcal{B})$ is the usual product of $\sigma -$algebra of sets.

It is clear from this definition that

$$i_{X \times Y} (A \times B) \equiv i_X (A) i_Y (B) , \forall A \in \mathcal{A} \text{ and } \forall B \in \mathcal{B}.$$

Thus, if several uncertain parameters in a problem are present with the uncertainty characterized by IVPMs, and all are independent, an IVPM for the product space can be formed by multiplication and subsequently used as an IVPM.

2. Interval-Valued Probability Optimization Interval-valued probability functions occurring as parameters to (6.1), (6.2) are considered next. Suppose we wish to solve the following,

$$opt\ z = f(x, \vec{a}),$$

$$\text{subject to: } g(x, \vec{b}) = 0,$$

where we assume that $\vec{a}$, and $\vec{b}$ are vectors of independent uncertain parameters, each with an associated IVPM. That is, $\hat{a} = \vec{a}$ and $\hat{b} = \vec{b}$. Assume the objective is maximize and the constraint can be violated at a cost so that the problem to be solved (dropping the vector notation) is

$$\max h\left(x, \hat{a}, \hat{b}\right) = f\left(x, \hat{a}\right) - p\left|g\left(x, \hat{b}\right)\right|, \qquad (6.14)$$

where p is a penalty vector.

The IVPM for the product space can be readily formed given the assumption that the uncertainty variables are independent. In this case, the IVPM, $i_{a \times b}$, is the joint distribution. The interval-valued expected value with respect to this IVPM is

$$\int_R h\left(\vec{x}, \hat{a}, \hat{b}\right) di_{a \times b}.$$

To optimize over such a value requires an ordering of intervals. One such ordering is to use the midpoint of the interval on the principle that in the absence of additional information, the midpoint is the best estimate for the true value. Another possible ordering is to use an interval risk/return multi-objective decision making similar to Markowitz's mean-variance models (see [43, 44]). For example, we can define functions

$$u : R^2 \to R$$

and

$$v : Int_R \to R^2$$

by setting, for any interval $I = [a, b]$, $v(I) = \left(\frac{a+b}{2}, b - a\right)$. Thus, v gives the midpoint and width of an interval. Then u would measure the preference for one interval over another considering both its midpoint and width (a risk measure). In this case, the optimization problem is

$$\max_x u\left(v\left(\int_R h(x, a, b)\, di_{axb}\right)\right).$$

Example 190 Consider the problem

$$\max f(\vec{x}, c) = 4x_1 + x_2$$
$$subject\ to$$
$$g_1(\vec{x}, \hat{a}_1, \hat{b}_1) = x_1 - [1, 3]x_2 + 4 = 0$$
$$g_2(\vec{x}, \hat{a}_2, \hat{b}_2) = \hat{2}x_1 - 5x_2 + 1 = 0$$
$$0 \le x_i \le 2$$

where $\hat{2} = 1/2/3/$, that is, $\hat{2}$ is the triangular number with support $[1, 3]$ and modal value at 2. For $\vec{p} = (1, 1)^T$,

$$h(\vec{x}, a, b) = 3x_1 + 6x_2 - \tilde{2}x_1 + [1, 3]x_2 - 5,$$

so that

$$\int_R h(\vec{x}, a, b) \, di_{axb}$$

$$= 3x_1 - \left[\int_0^1 (\alpha - 3)d\alpha, \int_0^1 (-\alpha - 1)d\alpha \right] x_1 + 6x_2 + [1, 3]x_2 - 5$$

$$= 3x_1 - \left[\frac{3}{2}, \frac{5}{2} \right] x_1 + 6x_2 + [1, 3]x_2 - 5$$

$$= \left[\frac{1}{2}, \frac{3}{2} \right] x_1 + [7, 9]x_2 - 5.$$

Note that -5 will not affect the optimization. It will be taken out of the optimization and then re-inserted at the end. Also note that we are maximizing the negative of the penalty, that is, we are minimizing the penalty. Continuing,

$$v \left(\int_R h(\vec{x}, a, b) \, di_{axb} \right) = (1, 1)x_1 + (8, 2)x_2$$

$$= (x_1 + 8x_2, x_1 + 2x_2).$$

As an example, let $u(y_1, y_2) = y_1 + y_2$. Then

$$\max_{x_i \in [0,2]} u \left(v \left(\int_R h(\vec{x}, a, b) \, di_{axb} \right) \right) = \max_{x_i \in [0,2]} \{2x_1 + 10x_2\}$$

$$= 24.$$

And putting the -5 back in, our solution is 19.

6.1.3.3 Penalty Method

The penalized approach given above can also be used for single distributions. Much of this section can also be found in [45]. Consider the following mathematical uncertainty optimization problem

$$\max z = f(\hat{c}, x) = \hat{c}^T x$$

$$\text{subject to: } g(\hat{a}, x) = \hat{A}x \leq \hat{b} \tag{6.15}$$

$$x \geq 0,$$

where the uncertainties in the components of A and/or b are possibility distributions. Now consider a penalty on violations of the following form

$$\hat{h}_i(\hat{A}, \hat{b}, x) = \max\left\{0, \hat{A}_i x - \hat{b}_i\right\},$$

where $\hat{A}_i$ is the ith row of $\hat{A}$. Let $d_i > 0$ be a cost per unit of violation penalty chosen a-priori. Note that $\hat{h}_i$ is a distribution where $\hat{h}$ is the corresponding vector-valued entity. Now form a new objective function

$$\hat{F}(x) = \hat{c}^T x - d^T \hat{h}.$$

Again, $\hat{F}$ is a generalized uncertainty entity. For generalized uncertainty types that are fuzzy intervals, $\hat{F}$ defines a fuzzy function. It is shown in [46] that the image at any vector x, $\hat{F}$, is a fuzzy interval. Thus, this fuzzy interval is completely characterized by its $\alpha - cuts$,

$$\hat{F}(x)_\alpha = \left(c^T x - d^T \max\{0, Ax - b\}\} \mid c, A, b \in \hat{c}_\alpha, \hat{A}_\alpha \text{ and } \hat{b}_\alpha \text{ respectively}\right).$$

This fuzzy number provides the possibility distribution for the outcome of taking action x. We wish to find the most favorable possibility distribution over all possible actions. Now that we have defined the distribution associated with the problem, we seek to determine a "best" solution. One reasonable approach is a minimax regret which is the subject of the next section. Another way is by using a type of expectation we call the *expected average* (see [47, 48]) of the fuzzy interval as the basis of comparing two fuzzy interval. The expected average provides the additional advantage of turning the range space into a Banach space for some applications (see [11]). Using the expected average point of view, the new optimization problem is as follows

$$\max EA(\hat{F}(x)) = EA(\hat{c}^T x - d^T \hat{h}),$$
$$\text{subject to: } x \geq 0.$$

Now,

$$\underline{F}_\alpha(x) = \hat{\underline{c}}_\alpha^T x - d^T \max(0, \overline{\hat{A}}_\alpha x - \overline{\hat{b}}_\alpha)$$
$$\overline{F}_\alpha(x) = \overline{\hat{c}}_\alpha^T x - d^T \max(0, \underline{\hat{A}}_\alpha x - \underline{\hat{b}}_\alpha)$$

where these two functions define the right and left endpoints of the α-cut of the fuzzy function evaluated at x where $\overline{F}_\alpha(x)$ is called the *optimistic* value of $\hat{F}(x)$ at possibility level α and $\underline{F}_\alpha(x)$ is called the *pessimistic* value of $\hat{F}(x)$. The modified problem now becomes

$$\max EA(\hat{F}(x)) = \frac{1}{2} \int_0^1 (\underline{F}_\alpha(x) + \overline{F}_\alpha(x)) d\alpha$$

$$= \frac{1}{2} \int_0^1 (\hat{\underline{c}}_\alpha^T x - d^T \max(0, \overline{\hat{A}}_\alpha x - \overline{\hat{b}}_\alpha) + \overline{\hat{c}}_\alpha^T x - d^T \max(0, \underline{\hat{A}}_\alpha x - \underline{\hat{b}}_\alpha)) d\alpha.$$

$$(6.16)$$

There is a closed form expression of (6.16) which is found in [49].

6.1.4 Minimum Maximum Regret

We next address the situation where there are uncertainties and with possible multiple objectives such as cost, environment impact, profit, risk, and so on, in a way that minimizes the regret one might have for making a decision without knowing what the actual state of nature is and later learning the true state of nature. Clearly, the natural aim is to find the best solution. In the multi-objective optimization problem under uncertainty, the natural question that arises is, "What is best?" in this context. If we are able to define "best", the next question is "How do we calculate 'best'?" What follows addresses these questions.

A classical formulation of a multi-objective optimization problem (MOOP) is:

$$\text{(MP) Minimize} \quad f(x)$$
$$s.t.: \quad g_i(x) \leqq b, \ i = 1, \ldots, m, \qquad (6.17)$$
$$x \in S,$$

where S is an open subset of $\mathbb{R}^n$, $f = (f_1, \ldots, f_p) : S \subseteq \mathbb{R}^n \to \mathbb{R}^p$ and $g = (g_1, \ldots, g_m) : S \subseteq \mathbb{R}^n \to \mathbb{R}^m$ are differentiable. To define what is the best solution for the MOOP, called the *efficient solution, sometimes also referred to as the Pareto optimal solution,* in the context of multi-objective optimization, the following notational conventions for equalities and inequalities are the following. If $x = (x_1, \ldots, x_n)$, $y = (y_1, \ldots, y_n) \in \mathbb{R}^n$, then

$$x = y \ \Leftrightarrow x_i = y_i, \quad \forall i = 1, \ldots, n,$$
$$x < y \ \Leftrightarrow x_i < y_i, \quad \forall i = 1, \ldots, n,$$
$$x \leqq y \ \Leftrightarrow x_i \leq y_i, \quad \forall i = 1, \ldots, n,$$
$$x \leq y \ \Leftrightarrow x_i \leq y_i, \quad \forall i = 1, \ldots, n, \quad \text{and there exist } j \text{ such that } x_j < y_j,$$
$$x \leq_j y \ \Leftrightarrow x_i \leq y_i \quad \forall i = 1, \ldots, n, \quad \text{and } x_j < y_j, \text{ for some } j \in \{1, \ldots, n\}.$$

We can now define what we mean by efficient solution, partial efficient solution and weakly efficient solution for the MOOP.

Definition 191 A feasible point, $\bar{x}$, is said to be an **efficient solution** for the MOOP if there does not exist another feasible point, x, such that $f(x) \leq f(\bar{x})$.

Now suppose there is uncertainty in the multi-objective problem, how do we proceed? Consider the following example:

Example 192 A person needs to travel from city A to city B. There are four options:
(1) car, (2) train, (3) airplane, (4) bus. It is necessary analyze two scenarios: (1) good
weather and (2) bad weather. The cost (in US\$) are $c_{11} = 120, c_{12} = 180, c_{21} = 150,$
$c_{22} = 160, c_{31} = 300, c_{32} = 310, c_{41} = 60, c_{42} = 80$ where c_{ij} is the cost of option
i in scenario j. The time of travel (in hours) are $t_{11} = 3, t_{12} = 5, t_{21} = 2.5, t_{22} = 4,$
$t_{31} = 1, t_{32} = 4, t_{41} = 4, t_{42} = 7$, where t_{ij} is the times of travel of the option i in the
scenario j.

A very useful approach, mainly when there is one single objective function, is
minimax regret, that is, find a solution that minimizes the worst regret. Some articles
in the literature can be found that use minimax regret solution for multi-objective
linear programming problems with interval objective functions [50–53]. However,
the minimax regret applied to multi-objective problem under uncertainty as presented
here is new.

6.1.4.1   Minimax Expected Regret Problem for the Multi-objective Case

This section defines the general bi-objective optimization problem under uncertainty.
The general multi-objective objective problem follows the same process as that which
is developed for the bi-objective case. The focus of this section is on the bi-objective
case for clarity.

Definition 193

$$\text{(BPU) Minimize } (f_1(x, \xi), f_2(x, \xi))$$
$$s.t.: \qquad g_i(x) \leqq b_i, \qquad i = 1, \ldots, m, \qquad (6.18)$$
$$x \in S, \xi \in U$$

where S is an open subset of $\mathbb{R}^n$, $U = (\xi_1, \ldots, \xi_n)$ is a finite set of all scenarios,
$f_i : S \times U \to, \mathbb{R} \, i = 1,2$, and $g = (g_1, \ldots, g_m) : S \times U \to \mathbb{R}^m$ are differentiable.

Denote, by $P(\xi)$, the bi-objective optimization problem for a fixed scenario ξ.
Abbreviate the minimax Regret Efficient solution by $MMRE$. Two definitions,
which are similar to what Kuhn et al. [54] presented, are given in the context of robust
bi-objective problems. However, the definition is tailored to the bi-objective minimax
regret problem (6.18). The generalization to k-*objective* functions is straight forward
albeit leading to more complex solution methods.

Definition 194 Solution $x \in X$ is a **weakly MMRE** if x is an efficient solution of
the $P(\xi)$ for at least one ξ in U.

Definition 195 Solution $x \in X$ is a **strongly MMRE** if x is an efficient solution of
the $P(\xi)$ for every ξ in U.

Clearly, if $x \in X$ is a strongly MMRE solution, then x is a weakly MMRE solution. Now, a new definition for minimax regret efficient solution for the bi-objective optimization under uncertainty problem is introduced. To do this, the MMRE solutions is taken in the context of minimax regret, which is defined next.

Definition 196 Solution $x \in X$ is **strictly MMRE** if x is an efficient solution of:

$$(\text{MMR-BPU})$$
$$\min_{x \in X} (\max_{\xi \in U}(f_1(x, \xi) - \min_x f_1(x, \xi)), \max_{\xi \in U}(f_2(x, \xi) - \min_x f_2(x, \xi)))$$

Theorem 197 *If $x^* \in X$ is a strongly MMRE solution then x^* is a strictly MMRE solution.*

Proof Let $x^* \in X$ be a strongly MMRE solution of the MMR-BPU, $z_1^* = \min_x f_1(x, \xi)$ and $z_2^* = \min_x f_2(x, \xi)$ and suppose that x^* is not a strictly MMRE solution. This means that there exists $\hat{x}$ such that:

$$(\max_{\xi \in U}(f_1(\hat{x}, \xi) - z_1^*(\xi)), \max_{\xi \in U}(f_2(\hat{x}, \xi) - z_2^*(\xi))$$
$$\leq (\max_{\xi \in U}(f_1(x^*, \xi) - z_1^*(\xi), \max_{\xi \in U}(f_2(x^*, \xi) - z_2^*(\xi)))$$

As a result there exists ξ_1, ξ_2, ξ_3 and ξ_4 in U (according to Definitions 1 and 5) such that:

$$(f_1(\hat{x}, \xi_1) - z_1^*, (f_2(\hat{x}, \xi_2) - z_2^*) \leq (f_1(x^*, \xi_3) - z*_1, (f_2(x^*, \xi_4) - z_2^*)$$

It is implies that:

1. $(f_1(\hat{x}, \xi_3) - z_1^*, (f_2(\hat{x}, \xi_4) - z_2^*) \leq (f_1(\hat{x}, \xi_1) - z_1^*, (f_2(\hat{x}, \xi_2) - z_2^*)$;

2. $(f_1(\hat{x}, \xi_3) - z_1^*, (f_2(\hat{x}, \xi_3) - z_4^*) \leq (f_1(\hat{x}, \xi_1) - z_1^*, (f_2(\hat{x}, \xi_2) - z_2^*)$;

3. $(f_1(\hat{x}, \xi_4) - z_1^*, (f_2(\hat{x}, \xi_4) - z_2^*) \leq (f_1(\hat{x}, \xi_1) - z_1^*, (f_2(\hat{x}, \xi_2) - z_2^*)$;

4. $(f_1(\hat{x}, \xi_4) - z_1^*, (f_2(\hat{x}, \xi_3) - z_2^*) \leq (f_1(\hat{x}, \xi_1) - z_1^*, (f_2(\hat{x}, \xi_2) - z_2^*)$.

So, we have:
$$(f_1(\hat{x}, \xi_3), f_2(\hat{x}, \xi_3)) \leq (f_1(x^*, \xi_3), f_2(x^*, \xi_3))$$

or

$$(f_1(\hat{x}, \xi_4), f_2(\hat{x}, \xi_4)) \leq (f_1(x^*, \xi_4), f_2(x^*, \xi_4)).$$

This is a contradiction, so x^* is an efficient solution for all scenarios. ∎

Now, Example 192 can be solved with respect to the point of view of our definitions.

Solving Example 192

It is easy see that for $x_1 =$ car, $x_2 =$ train, $x_3 =$ airplane, $x_4 =$ bus:

- x_2, x_3, x_4 are weakly MMRE solutions, because they are efficient solutions for some scenario.
- x_2, x_4 are strongly MMRE solutions, because both are efficient solutions for all scenarios.

To find the strictly MMRE solutions, it is necessary to analyze the efficiency of the following vectors: $fm(x_1) = (120, 4)$, $fm(x_2) = (100, 3)$, $fm(x_3) = (250, 3)$ and $fm(x_4) = (20, 6)$, where:

$$fm(x_i) = (\max_{\xi \in U}(f_1(x, \xi) - \min_x f_1(x, \xi)), \max_{\xi \in U}(f_2(x, \xi) - \min_x f_2(x, \xi))).$$

Therefore, we conclude that x_2, x_3 and x_4 are MMRE solutions (x_1 is not a strictly MMRE solution because $fm(x_2) < fm(x_1)$).

6.1.4.2   The Linear Bi-objective Minimax Regret Optimization Problem with Generalized Uncertainty

Consider the following classic example of a linear optimization problem under uncertainty (see [55, 56]).

Example 198 A farmer raises wheat, corn, and sugar beets on 500 acres of land. The planting cost per acre of wheat, corn, and sugar beet are $150, $230, and $260, respectively. The farmer knows that at least 200 tons of wheat and 240 tons of corn are needed for cattle feed. These amounts can be raised on the farm or bought from a market. Any production in excess of the feeding requirement will be sold. Selling prices are $170 and $150 per ton of wheat and corn, respectively. The purchase price is higher, $238 per ton of wheat and $210 per ton of corn. Another possible crop is sugar beet, which sells at $36 per ton. However, the state commission imposes a quota on sugar beet production of 6,000 tons. Any amount in excess of the quota can be sold only at $10 per ton. The mean yield on the farmer's land is 2.5, 3, and 20 tons per acre for wheat, corn, and sugar beets, respectively. Now, suppose that: *The farmer collects weather data for the past 360 years of his town from local government sources, and finds out that there were 120 years of bad weather, but the farmer could not tell what the pattern of this bad weather was from the data. There are 180 years, in which no bad things happen; no monster storms, no devastating hail, no outbreak of insects. From the weather data, the farmer can determine the planting schedule during these 180 years. However, because there are no data on yields, the farmer does not know for sure that these 180 years have above average or average yields. Moreover, the weather data shows that there are 60 years of chaotic conditions; mixed good and bad seasons for farming. The farmer cannot tell which of these 60 years provide below average, average, or above average crops yields. For this information, the farmer has a random set of yields with respect to the uncertainty*

$$\hat{v} = v_1 = [2, 2.4, 16]^T, v_2 = [2.5, 3, 20]^T, v_3 = [3, 3.6, 24]^T$$

such that:

$$m(\{v_1\}) = 1/3, m(\{v_2, v_3\}) = 1/2, m(\{v_1, v_2, v_3\}) = 1/6,$$

where v_1, v_2, and v_3 refer to the vectors of the below average, average, and above average yields of wheat, corn, and sugar beets.

The farmer also wishes minimize the quantity of water used because of the environmental impact. For each acre of wheat/beet/corn, the consumption is proportional to 7.25/6.5/10 m^3 of water.

Before solving this example problem, some more concepts will be defined. A problem is a bi-objective linear optimization problem under generalized uncertainty (BLPGU), if it satisfies the following definition.

Definition 199

$$\text{(BLPGU) Minimize } (\max_{p \in M}(z_1(x, y) - \min_{x,y} z_1(x, y)), f_2(x))$$
$$s.t. : \qquad Ax \leqq b_i, i = 1, \ldots, m,$$
$$p \in M$$

where $z_1(x, y) = c^t x + q_1 \sum_{k=1}^{K_1} p_1^k y_1^k + q_2 \sum_{k=1}^{K_2} p_2^k y_2^k + \cdots + q_m \sum_{k=1}^{K_m} p_m^k y_m^k$, $M = (p_1, \ldots, p_n)$ is a finite set of probabilities of all realizations of all scenarios $U = (\xi_1, \ldots, \xi_m)$, $f_1 : S \times U \to \mathbb{R}$, and $f_2 : S \to \mathbb{R}$ are linear functions.

We are now in a position to define what we mean by an efficient minimax regret solution for the bi-objective case.

Definition 200 Solution $x \in X$ is strictly MMRE for BLPGU if x is an efficient solution (according to Definitions 194, 195, or 196) of the BLPGU.

Example 198 is a BLPGU. Thus, the problem can be mathematically formulated in the following way:

$$\text{(MP-Ex2)}$$
$$\min_x \qquad (\max_{p \in M}((z_1(x) - \min_x z_1(x, p), (f_2(x))))$$
$$s.t. : \qquad AX \leqq b,$$
$$p \in M$$

where:

$$z_1(X) = 150x_1 + 230x_2 + 260x_3 +$$
$$p_1 * 238x_4 - p_1 * 170x_5 + p_2 * 238x_6 - p_2 * 170x_7 + p_3 * 238x_8 +$$
$$- p_3 * 170x_9 + p_1 * 210x_{10} - p_1 * 150x_{11} + p_2 * 210x_{12} - p_2 * 150x_{13} +$$
$$+ p_3 * 210x_{14} - p_3 * 150x_{15} - p_1 * 36x_{16} - p_1 * 10x_{17} +$$

$$- p_2 * 36x_{18} - p_2 * 10x_{19} - p_3 * 36x_{20} - p_3 * 10x_{21}$$
$$f_2(x) = 7.250_1 + 6.5x_2 + 10x_3.$$

The resulting coefficient matrix, right-hand side vector and bounds are:

$$A = \begin{bmatrix}
1 & 1 & 1 & 0 & 0 & 0 & 0 & 0 & 0 & 0 & 0 & 0 & 0 & 0 & 0 & 0 & 0 & 0 & 0 & 0 & 0 \\
7.25 & 6.5 & 10 & 0 & 0 & 0 & 0 & 0 & 0 & 0 & 0 & 0 & 0 & 0 & 0 & 0 & 0 & 0 & 0 & 0 & 0 \\
-2 & 0 & 0 & -1 & 1 & 0 & 0 & 0 & 0 & 0 & 0 & 0 & 0 & 0 & 0 & 0 & 0 & 0 & 0 & 0 & 0 \\
-2.5 & 0 & 0 & 0 & 0 & -1 & 1 & 0 & 0 & 0 & 0 & 0 & 0 & 0 & 0 & 0 & 0 & 0 & 0 & 0 & 0 \\
-3 & 0 & 0 & 0 & 0 & 0 & 0 & -1 & 1 & 0 & 0 & 0 & 0 & 0 & 0 & 0 & 0 & 0 & 0 & 0 & 0 \\
0 & -2.4 & 0 & 0 & 0 & 0 & 0 & 0 & 0 & -1 & 1 & 0 & 0 & 0 & 0 & 0 & 0 & 0 & 0 & 0 & 0 \\
0 & -3 & 0 & 0 & 0 & 0 & 0 & 0 & 0 & 0 & -1 & 1 & 0 & 0 & 0 & 0 & 0 & 0 & 0 & 0 & 0 \\
0 & -3.6 & 0 & 0 & 0 & 0 & 0 & 0 & 0 & 0 & 0 & 0 & -1 & 1 & 0 & 0 & 0 & 0 & 0 & 0 & 0 \\
0 & 0 & -16 & 0 & 0 & 0 & 0 & 0 & 0 & 0 & 0 & 0 & 0 & 0 & 0 & 1 & 1 & 0 & 0 & 0 & 0 \\
0 & 0 & -20 & 0 & 0 & 0 & 0 & 0 & 0 & 0 & 0 & 0 & 0 & 0 & 0 & 0 & 0 & 1 & 1 & 0 & 0 \\
0 & 0 & -24 & 0 & 0 & 0 & 0 & 0 & 0 & 0 & 0 & 0 & 0 & 0 & 0 & 0 & 0 & 0 & 0 & 1 & 1
\end{bmatrix}$$

$$b = (500, -200, -200, -200, -240, -240, -240, 0, 0, 0)$$
$$x_{16} \le 6000$$
$$x_{18} \le 6000$$
$$x_{20} \le 6000$$
$$x_j \ge 0 \text{ for all } j = 1, \ldots, 21.$$

For this example x_1 is the acreage of land devoted to wheat, x_2 is the acreage of land devoted to corn, x_3 is the acreage of land devoted to sugar beets, $x_4/x_6/x_8$ are the tons of wheat purchased in the scenario $1/2/3$, $x_5/x_7/x_9$ are the tons of corn purchased in the scenario $1/2/3$, $x_{10}/x_{12}/x_{14}$ are the tons of wheat sold in the scenario $1/2/3$, $x_{11}/x_{13}/x_{15}$ are the tons of corn sold in the scenario $1/2/3$, $x_{16}/x_{18}/x_{20}$ are the tons of sugar beets sold at US\$36 in the scenario $1/2/3$, $x_{17}/x_{19}/x_{21}$=tons of sugar beets sold at US\$10 in the scenario $1/2/3$, and ρ_1, ρ_2, ρ_3 represent the probability information about the weather that generates the set $M = \{p_1^k, p_2^k, p_3^k\}, k = 1, \ldots, L$ that will be calculated in the next section.

6.1.4.3   The H-ϵ Heuristic Applied to the MiniMax Regret Under Uncertainty

This section presents a heuristic based on the ϵ-constraint method, that we will call of the H-ϵ method, a modification of [57].

Definition 201 Let the BLPGU be given by:

(BLPGU) Minimize $(\max_{p \in M}(z_1(x, y) - \min_{x, y} z_1(x, y)), f_2(x))$
$$\begin{array}{lll} s.t.: & Ax \le b_i, & i = 1, \ldots, m, \\ & p \in M \end{array}$$

where

$$z_1(x, y) = c^t x + q_1 \sum_{k=1}^{K_1} p_1^k y_1^k + q_2 \sum_{k=1}^{K_2} p_2^k y_2^k + \cdots + q_m \sum_{k=1}^{K_m} p_m^k y_m^k, \quad M = (p_1, \ldots, p_n)$$

is a finite set of probabilities of all realizations of all scenarios

$$U = (\xi_1, \ldots, \xi_m),$$
$$f_1 : S \times U \to \mathbb{R} \text{ and } f_2 : S \to \mathbb{R}$$

are linear functions. We associate this problem with a family of linear optimization problems $P_\epsilon, \epsilon \in \mathbb{R}$:

$$P_\epsilon \text{ Minimize } h(x) = (\max_{\xi \in U}((z_1(x, y) - \min_{x,y} z_1(x, y))))$$
$$s.t. : \qquad\qquad Ax \leq b_i, \qquad\qquad i = 1, \ldots, m,$$
$$f_2(x) \leq \epsilon$$
$$p \in M$$

Definition 202 x^* is a possibly MMRE solution if it is an optimum solution for problem $P_\epsilon, \epsilon \in R$.

Theorem 203 *Let x^* be a strictly MMRE solution for BPLGU. Then x^* is a possible MMRE solution for BPLGU, where $\epsilon = f_2(x^*)$.*

Proof Let x^* be a strictly MMRE solution for BPLGU. Suppose that there exist a feasible solution $\hat{x}$ for BPLU such that $h(\hat{x}) < h(x^*)$. Therefore,

$$(h(\hat{x}), f_2(\hat{x})) \leq (h(x^*), f_2(x)),$$

which is a contradiction. ∎

Remark 204 The contrapositive of this theorem is not valid. For example, consider that in Example 198 we have uncertainty just in the cost, $f_1(x)$. Rewrite the BLPGU problem in the follow way: A person needs travel from city A to city B. There are four options: (1) car, (2) train, (3) airplane, (4) bus. It is necessary to analyze two scenarios: (1) good weather and (2) bad weather. The cost (in US$) is $c_{11} = \$120$, $c_{12} = \$180$, $c_{21} = \$150$, $c_{22} = 160$, $c_{31} = \$300$, $c_{32} = \$310$, $c_{41} = \$60$, $c_{42} = \$80$ {where c_{ij} is the cost of option i in scenario j. The travel time (in hours) is $t_1 = 3$, $t_2 = 2.5$, $t_3 = 1$, $t_4 = 4$. Let $f_2(x)$ be the function time of travel, and $\epsilon = 3$. In this case we have that $x_1\$$ is the optimum solution. However, x_1 is not a MMRE solution.

Remark 205 Is it possible increase the feasible region and have a worse solution? For the minimax the answer is yes. Consider that in the example above we have a new way to travel from city A to city B, a fifth option (5), an express train with $c_{51} = \$30$, $c_{52} = \$170$ and $t_5 = 4$. For this case, the uncertainty is just in the cost, which we will denote $f_1(x)$. Rewrite the problem in the follow way: A person needs travel from city A to city B. There are four options: (1) car, (2) train, (3) airplane, (4) bus. It is necessary to analysis two scenarios: (1) good weather and (2) bad weather. The cost (in US \$) is $c_{11} = \$120$, $c_{12} = \$180$, $c_{21} = \$150$, $c_{22} = 160$, $c_{31} = \$300$, $c_{32} = \$310$, $c_{41} = \$60$, $c_{42} = \$80$ {where c_{ij} is the cost of option i in scenario j. The travel time (in hours) is $t_1 = 3$, $t_2 = 2.5$, $t_3 = 1$, $t_4 = 4$. This case results in MMRE solutions x_1, x_2, x_3, x_4 all with bigger $f_1(x)$ than before the introduction of the option x_5.

We can now propose the H-ϵ method to find minimax regret efficient solution for the bi-objective linear optimization problem under uncertainty. This method joins ideas of ϵ-constraint algorithm (see [57]) and the algorithm proposed in [56], but the algorithm does not guaranteed that all the efficient solutions found are MMRE, only that some of the efficient solutions are found.

Algorithm 206 (H) Given: $f_1(x; p)$, $f_2(x)$, A, b, p—the probability measure vector, $\epsilon = (\epsilon_1, \ldots, \epsilon_T)$, $k = 1, \ldots, T$.

Step1: Initialization. Choose $p(1) = p$, Solve $f_1(x)$ and obtain an optimal solution $x(1)$.

Step 2: Set $j = 1$.

Step 3: Solve the following current relaxed problem to obtain an optimal solution $R(j)$; $x(j)$:

$$\min_{R,x} R$$

$$st\ R \geq 0$$

$$R \geq f_{1p(i)}(x) - f_{1p(i)}(x_j), i = 1, \ldots, j$$

$$f_2(x) \leq \epsilon_k$$

Step 4: Start with $p^{(j)}$ and for i not in $p^{(r)*}$ like [56]. Calculate $F(j)_{p^{(r)*}}$ and its optimal solution $x(j+1)$.

Step 5: If $F(j)_{p^{(r)*}} > R(j)$ set $p^r = p^{j+1}$. Set $j = j+1$, return to step 2.

Step 6: If $F(j)_{p^{(r)*}} > R(j)$ select p that has not been used in this iteration of step 3 and reprocess the system as proposed in [56] until we find a $p^{(r)}$ such that $F(j)_{p^{(r)*}} > R(j)$. Then go to step 4, $M = \varnothing$ and we terminate the procedure. An efficient solution to the BLPU is x_j.

We next use this method to obtain an approximation of the Pareto curve for Example 198.

Solving Example 198: To obtain the probabilistic information ρ_1, ρ_2, ρ_3 we use the same procedure as [56]. For example, the probability mass functions that generate the smallest expected values are calculated in the follow way:

$$p_1^1 = Nec(v_1, v_2, v_3) - Nec(v_2, v_3) = 1/2$$

$$p_2^1 = Nec(v_2, v_3) - Nec(v_3) = 1/2$$

Table 6.1 Solutions obtained using the H-ϵ method

$\min_x \{\max_{\xi \in U}(f_1(x, \xi) - \min_x f_1(x))\}$	$f_2(x)$
977.289	2000
977.260	2150
977.219	2300
977.014	2450
976.532	2600
976.516	2750
976.453	2900
976.348	3050
976.167	3350
910.456	3500
599.567	3650
211.805	3800

$$p_3^1 = Nec(v_3) = 0$$

where Nec is the necessity measure [56]. Recall that the Nec measure was defined in Chaps. 2 and 3 and would be used here. To generate the approximation of the Pareto curve, 21 values of ϵ were used, $\epsilon_i = 2000 + 150 * i, i = 0, \ldots, 20$. and the values obtained are shown in Table 6.1.

Remark 207 It is important to recall that the solutions presented in the Table 6.1 can not be a MMRE solutions.

6.2 Example—Investment Strategies

Four methods are illustrated by an example that considers three possible investments (I_1, I_2, I_3), which can each end up in three different scenarios optimistic, medium, pessimistic. Each investment can have the return (in percentage) shown in the Tables 6.2–6.6.

Let us now consider four ways to select the best investment:

Table 6.2 Return

Investments	Optimistic	Medium	Pessimistic
I_1	15	3	−10
I_2	10	5	0
I_3	7	6	3

Table 6.3 MAXIMAX criterion

Investments	Optimistic	Medium	Pessimistic	Max
I_1	15	3	-10	**15**
I_2	10	5	0	10
I_3	7	6	3	7

Table 6.4 MAXIMIN criterion

Investments	Optimistic	Medium	Pessimistic	Max
I_1	15	3	-10	-10
I_2	10	5	0	0
I_3	7	6	3	**3**

Table 6.5 Laplace criterion

Investments	Optimistic	Medium	Pessimistic	Average
I_1	15	3	-10	8/3
I_2	10	5	0	15/3
I_3	7	6	3	**16/3**

Table 6.6 MINIMAX regret criterion

Investments	Optimistic	Medium	Pessimistic	Average
I_1	0	2	13	13
I_2	5	1	3	**5**
I_3	8	0	0	8

1. **MAXIMAX**: This option maximizes the maximum return available, without risk aversion, that is, the best of the best. So, in this example we select the maximum return considering all scenarios: So, in this example, investment I_1 = Table 6.3 is selected:
2. **MAXIMIN**: This option maximizes the minimum return for each scenario, that is, the best of the worst. So, in this example we select the maximum minimum return considering all scenarios: So in this example, investment I_3 = Table 6.4 is selected:
3. **Laplace**: This option maximizes the return average for each scenario, that is, the best of the average. So, in this example, investment I_3 = Table 6.5 is selected:
4. **MINIMAX Regret**: This option minimizes the maximum regret, that is, the best investment considering the maximum regret. So, in this example, investment I_2, whose average is 3, is selected:

6.2.1   MiniMax Regret Under Generalized Uncertainty

This section revisits minimax regret, but from a generalized uncertainty point of view. The concept of minimum of maximum (minimax) regret is a well known approach in making a decision based on many alternatives that could be chosen. In the context of generalized uncertainty optimization, a generalized uncertainty entity is a set of distributions that contains the unknown distribution of our problem. All that is known with certainty is that the true distribution is bracketed between the given lower and upper cumulative probabilities. The uncertainty we consider in this section are discrete random sets. Fuzzy intervals with a possibility/necessity semantic, and plausibility and belief are uncertainties that can be transformed to random sets. Given a set of n realizations of an uncertainty $\hat{u}$ as $\{u_1, u_2, \ldots, u_n\}$, where $u_1 \leq u_2 \leq \cdots \leq u_n$, Nguyen [58] proved that the lowest and largest expected values of $\hat{u}$ can be evaluated by using the following density functions $\underline{f}$ and $\overline{f}$, where, recalling our development of belief/plausibility of Demster [59] and $\overline{\text{Shafer}}$ [60, 61],

$$\underline{f}(u_k) = Bel\left(\{u_k, u_{k+1}, \ldots, u_n\}\right) - Bel\left(\{u_{k+1}, u_{k+2}, \ldots, u_n\}\right) \qquad (6.19)$$

and

$$\overline{f}(u_k) = Bel\left(\{u_1, u_2, \ldots, u_k\}\right) - Bel\left(\{u_1, u_2, \ldots, u_{k-1}\}\right). \qquad (6.20)$$

Our general linear programming problem with uncertainty is

$$z = \min c^T x$$

$$\text{Subject to: } \hat{A}x \geq \hat{b}, \quad x \geq 0. \qquad (6.21)$$

Based on the concept of an expected recourse model in stochastic programming (see, for example, [55]), we could use the bounds (6.19), (6.20), $\underline{f}$ and $\overline{f}$, to generate an optimistic and pessimistic expected recourse problems for system (6.21). Here, we are only considering the discrete pairwise independent case for the generalized uncertainties in the linear optimization model. The optimistic expected recourse problem is the minimum of the expected recourse values over a set M of probability vector measures associated with the uncertainties in (6.21). Note that in order to get the optimistic and pessimistic solutions, we need only $\underline{f}$ and $\overline{f}$. We can also reduce the size M to a finite set of all possible $\underline{f}$ and $\overline{f}$ as shown by the example and table that follows.

An element in the set M is a vector of probability $f = [f_1, f_2, \ldots, f_m]$, where m is the number of constraints in (6.21). Given the pairwise independence assumption, let $f_i = \left[f_i^{(1)}, f_i^{(2)}, \ldots, f_i^{(K_i)}\right]$ be a joint probability vector for the ith constraint, that is, $f_i^{(k)}$ is a joint probability for the kth scenario, $k = 1, 2, \ldots, K_i$ where $\sum_{k=1}^{K_i} f_i^{(k)} = 1$ and let the vector $y = [y_1, y_2, \ldots, y_m]$ be the recourse action of scenarios for each constraints with $y_i = \left[y_i^{(1)}, y_i^{(2)}, \ldots, y_i^{(K_i)}\right]$. Then, the following theorem translates optimistic and pessimistic expected recourse problems as linear programs. The proof of the theorem could be found in [8, 62].

Theorem 208 (see [8, 62]) *Let q_i be the penalty price of violating constraint i of (6.21). Assume that all uncertainties are pairwise independent from each other. Given*

$$z_f(x, y) = c^T x + q_1 \sum_{k=1}^{K_1} f_1^{(k)} y_1^{(k)} + a_2 \sum_{k=1}^{K_2} f_2^{(k)} y_2^{(k)} + \cdots + q_m \sum_{k=1}^{K_m} f_m^{(k)} y_m^{(k)},$$

suppose that $\min_{x,y} z_f(x, y)$ is bounded for each $f \in M$. Then there exist constructible $\underline{f}$ and $\overline{f} \in M$ such that

$$\min_{f \in M} \min_{x,y} z_f(x, y) = c^T x + q_1 \sum_{k=1}^{K_1} \underline{f}_1^{(k)} y_1^{(k)} + a_2 \sum_{k=1}^{K_2} \underline{f}_2^{(k)} y_2^{(k)} + \cdots + q_m \sum_{k=1}^{K_m} \underline{f}_m^{(k)} y_m^{(k)},$$

$$\max_{f \in M} \min_{x,y} z_f(x, y) = c^T x + q_1 \sum_{k=1}^{K_1} \overline{f}_1^{(k)} y_1^{(k)} + a_2 \sum_{k=1}^{K_2} \overline{f}_2^{(k)} y_2^{(k)} + \cdots + q_m \sum_{k=1}^{K_m} \overline{f}_m^{(k)} y_m^{(k)}.$$

We do not know the exact probability distribution of (6.21) so that all probabilities in M are considered as the alternatives in a minimum maximum expected regret approach. That is, we would like to know the best of the worst regret over $f \in M$. Concretely, the minimum maximum regret is given by

$$\min_{x,y} \max_{f \in M} \left(z_f(x, y) - \min_{x,y} z_f(x, y) \right). \tag{6.22}$$

A standard procedure to deal with a minimax regret problem is to do the relaxation procedure by increasing the constraint of the relaxed problem in each iteration until reaching some termination criteria.

Algorithm 209 Relaxation procedure

1. Initialization. Choose $f^1 = \underline{f}$. Solve $\min_{x,y} z_{\underline{f}}(x, y)$ and obtain an optimal solution (w^1, v^1). Set $j = 1$.
2. Solve the following current relaxed problem to obtain an optimal solution $\left(R^j, (x^j, y^j) \right)$.

$$\min_{R,(x,y)} R$$

Subject to: $R \geq 0$

$$R \geq z_{f^i}(x, y) - z_{f^i}(w^i, v^i), \quad i = 1, 2, \ldots, j.$$

3. Obtain an optimal solution $\left(f^{j+1}, (w^{j+1}, v^{j+1}) \right)$ where its optimal value Z^j is

$$Z^j = \max_{f \in M, (x,y)} \left(z_f(x^j, y^j), z_f(x, y) \right).$$

4. If $Z^j \le R^j$, terminate the procedure. An optimal solution to the minimax expected regret model (6.22) is (x^j, y^j). Otherwise, set $j = j + 1$ then return to step 2. $\square$

Step 3 of the relaxation procedure is the most complicated. A modified relaxation problem developed by Thipwiwatpotjana and Lodwick [8], is applied to the upper and lower probability functions of the random set to speed up the calculation of step 3 in the original procedure. In particular, given a probability f, there is a sequence of probabilities generated upon the order of $y^j - y$ through the following relationship.

$$
\left.
\begin{aligned}
& Z_f^j \\
&= \max_{(x,y)} \left(z_f(x^j, y^j) - z_f(x, y) \right) \\
&= \max_{(x,y)} \left(c^T(x^j - x) + q_1 \sum_{k=1}^{K_1} f_1^{(k)}(y_1^{j^{(k)}} - y_1^{(k)}) + \cdots + q_m \sum_{k=1}^{K_m} f_m^{(k)}(y_m^{j^{(k)}} - y_m^{(k)}) \right) \\
&= c^T(x^j - x_f) + q_1 \sum_{k=1}^{K_1} f_1^{(k)}(y_1^{j^{(k)}} - y_{1_f}^{(k)}) + \cdots + q_m \sum_{k=1}^{K_m} f_m^{(k)}(y_m^{j^{(k)}} - y_{m_f}^{(k)}) \\
&\le c^T(x^j - x_f) + q_1 \sum_{k=1}^{K_1} f_1^{r(k)}(y_1^{j^{(k)}} - y_{1_f}^{(k)}) + \cdots + q_m \sum_{k=1}^{K_m} f_m^{r(k)}(y_m^{j^{(k)}} - y_{m_f}^{(k)}) \\
& \text{where } f_i^r, \text{ corresponding to the order of}(y_i^{j^{(k)}} - y_{i_f}^{(k)}), \text{ is obtained} \\
& \text{by applying the system (6.20)} \\
&\le c^T(x^j - x_{f^r}) + q_1 \sum_{k=1}^{K_1} f_1^{r(k)}(y_1^{j^{(k)}} - y_{1_{f^r}}^{(k)}) + \cdots + \\
& q_m \sum_{k=1}^{K_m} f_m^{r(k)}(y_m^{j^{(k)}} - y_{m_{f^r}}^{(k)}) \\
&= \max_{(x,y)} \left(z_{f^r}(x^j, y^j) - z_{f^r}(x, y) \right) \\
&= Z_{f^r}^j.
\end{aligned}
\right\}
$$

$$(6.23)$$

The procedure continues to find a new f^r using the above relation (6.23) until the new probability f^{r*} can be applied to the each suborder of $(y_i^{j^{(k)}} - y_{i_{f^{r*}}}^{(k)})$. It is clear that $Z_{f^{r*}}^j \le Z^j$. However, if $Z_{f^{r*}}^j > R^k$, then quit step 3 even if Z^j is not computed and return to step 2 in the relaxation procedure with $j = j + 1$ to find an updated optimal solution $\left(R^j, (x^j, y^j) \right)$. Otherwise, select f that has not been used in this iteration of step 3 and reprocess the system (6.23) until f^{r*} is found such that $Z_{f^{r*}}^j > R^k$ and obtain an optimal solution to the minimax regret problem. Hence, these new steps alter step 3 and a modified version of the relaxation procedure is the following.

Modified Relaxation Procedure

1. Initialization. Choose $f^1 = \underline{f}$. Solve $\min\limits_{(x,y)} z_{\underline{f}}(x, y)$, and obtain an optimal solution (w^1, v^1). Set $j = 1$.
2. Solve the following current relaxed problem to obtain an optimal solution $\left(R^j, (x^j, y^j)\right)$.

$$\min_{R,\,(x,y)} \quad R$$
$$\text{Subject to: } R \geq 0$$
$$R \geq z_{f^i}(x, y) - z_{f^i}(w^i, v^i), \quad i = 1, 2, \ldots, j.$$

3. Start with f^j and work on the system (6.23) to find f^{r*}. Calculate $Z^j_{f_{r*}}$ and its optimal solution (w^{j+1}, v^{j+1}).
4. If $Z^j_{f_{r*}} > R^j$, set f^{r*} as f^{j+1}. Set $j = j + 1$, then return to step 2.
5. If $Z^j_{f_{r*}} \leq R^j$, select f that has not been used in this iteration of step 3, and reprocess the system (6.23) until we find f^{r*} such that $Z^j_{f_{r*}} > R^k$, then continue step 4. Otherwise, we terminate the procedure. An optimal solution to the minimax expected regret model (6.22) is $\left(x^j, y^j\right)$.   ☐

The next example indicates how to compute the generalized uncertainty minimax regret problem.

Example 210 A linear program P_1 with uncertain vector $(\hat{a}_{11}, \hat{a}_{12}, \hat{b}_1)$

$$P_1 : min \qquad 3x_1 + 2x_2$$
$$\text{Subject to: } \hat{a}_{11}x_1 + \hat{a}_{12}x_2 \geq \hat{b}_1$$
$$x_1, x_2 \geq 0$$

Suppose we know realizations of $(\hat{a}_{11}, \hat{a}_{12}, \hat{b}_1)$ and their probabilities as :
Below average(realization1) : $(a_{11}^{(1)}, a_{12}^{(1)}, b_1^{(1)}) = (7, 8, 45)$, with probability $f_1^{(1)}$,
   Average(realization2) : $(a_{11}^{(2)}, a_{12}^{(2)}, b_1^{(2)}) = (8, 9, 50)$, with probability $f_1^{(2)}$ and
   Above average(realization3) : $(a_{11}^{(3)}, a_{12}^{(3)}, b_1^{(3)}) = (9, 10, 53)$, with probability $f_1^{(3)}$.
   The corresponding stochastic expected recourse problem with recourse variables of the demand constraints w_1, w_2, w_3 is

$$P_2 : min \quad 3x_1 + 2x_2 + s_1 f_1^{(1)} w_1^{(1)} + s_1 f_1^{(2)} w_1^{(2)} + s_1 f_1^{(3)} w_1^{(3)}$$

$$\text{Subject to: } \begin{array}{ll} 7x_1 + 8x_2 + w_1^{(1)} & \geq 45 \\ 8x_1 + 9x_2 + w_1^{(2)} & \geq 50 \\ 9x_1 + 10x_2 + w_1^{(3)} & \geq 53 \\ x_1, x_2, w_1^{(1)}, w_1^{(2)}, w_1^{(3)} & \geq 0. \end{array} \qquad (6.24)$$

How do we interpret a result of P_2 if we do not know the exact probability, for example, if $f_1^{(1)} \in [\frac{1}{3}, \frac{1}{2}]$, $f_1^{(2)} \in [\frac{1}{6}, \frac{2}{3}]$ and $f_1^{(3)} \in [\frac{1}{6}, \frac{1}{2}]$? One of the approaches is to use an idea of interval expected value. Therefore, to find an interval expected value to the objective function of P_2 with unknown probability $(f_1^{(1)}, f_1^{(2)}, f_1^{(3)})$ we have to know the ordering of $w_1^{(1)}, w_1^{(2)}, w_1^{(3)}$. All possible ordering cases of $w_1^{(1)}, w_1^{(2)}, w_1^{(3)}$ are in the following table. Hence, the set M of probability vector measures associated with uncertainty $(\hat{a}_{11}, \hat{a}_{12}, \hat{b}_1)$ can be reduced to only 4 probabilities as $(f_{1_1}, f_{1_2}, f_{1_3}) = (\frac{1}{2}, \frac{1}{3}, \frac{1}{6})$, $(\frac{1}{2}, \frac{1}{6}, \frac{1}{3})$, $(\frac{1}{3}, \frac{1}{2}, \frac{1}{6})$ and $(\frac{1}{3}, \frac{1}{6}, \frac{1}{2})$. After applying modified relaxation procedure, the result of the minimax regret approach to P_1 is

$$R = 0, \ x_1 = 0, \ x_2 = 5.625, \ w_1^{(1)} = 0, \ w_1^{(2)} = 0 \ \text{and} \ w_1^{(3)} = 0.$$

This generates the following.

Ordering cases	Lower prob $\mathbf{f}_i$			Upper prob $\overline{\mathbf{f}}_i$		
	$f_1^{(1)}$	$f_1^{(2)}$	$f_1^{(3)}$	$\overline{f}_1^{(1)}$	$\overline{f}_1^{(2)}$	$\overline{f}_1^{(3)}$
$w_1^{(1)} \le w_1^{(2)} \le w_1^{(3)}$	$\frac{1}{2}$	$\frac{1}{3}$	$\frac{1}{6}$	$\frac{1}{3}$	$\frac{1}{6}$	$\frac{1}{2}$
$w_1^{(1)} \le w_1^{(3)} \le w_1^{(2)}$	$\frac{1}{2}$	$\frac{1}{6}$	$\frac{1}{3}$	$\frac{1}{3}$	$\frac{1}{6}$	$\frac{1}{2}$
$w_1^{(2)} \le w_1^{(1)} \le w_1^{(3)}$	$\frac{1}{3}$	$\frac{1}{2}$	$\frac{1}{6}$	$\frac{1}{3}$	$\frac{1}{2}$	$\frac{1}{6}$
$w_1^{(2)} \le w_1^{(3)} \le w_1^{(1)}$	$\frac{1}{3}$	$\frac{1}{2}$	$\frac{1}{6}$	$\frac{1}{3}$	$\frac{1}{2}$	$\frac{1}{6}$
$w_1^{(3)} \le w_1^{(1)} \le w_1^{(2)}$	$\frac{1}{3}$	$\frac{1}{6}$	$\frac{1}{2}$	$\frac{1}{2}$	$\frac{1}{6}$	$\frac{1}{3}$
$w_1^{(2)} \le w_1^{(2)} \le w_1^{(1)}$	$\frac{1}{3}$	$\frac{1}{6}$	$\frac{1}{2}$	$\frac{1}{2}$	$\frac{1}{3}$	$\frac{1}{2}$

6.3 Summary

This chapter presented five different ways that optimization under generalized uncertainty can be carried out. A study about which methods to use depending on the problem was discussed in [63].

6.4 Exercises

Exercise 211 Provide an example that an optimal solution to a minimax regret problem is not one of the optimal solutions of realizations.

Exercise 212 Given $U = \{u_1, u_2, u_3\}$, where $pr(u_1) \in [1/3, 1/2]$, $pr(u_2) \in [1/6, 2/3]$, $pr(u_3) \in [1/4, 1/2]$, verify that the interval range of probability represents belief and plausibility or not. Moreover if $u_1 = 1, u_2 = 2, u_3 = 3$, find the minimum and maximum expected values.

Exercise 213 Generate the corresponding stochastic expected recourse model and minimax regret model (when not considering probabilities) for the following linear programming with uncertainty.

$$
\begin{aligned}
\min \quad & 2x_1 \quad +3x_2 \\
\text{s.t.} \quad & \hat{a}_{11}x_1 \quad +6x_2 \geq \hat{b}_1 \\
& 3x_1 +\hat{a}_{22}x_2 \geq \hat{b}_2 \\
& x_1, \qquad x_2 \geq 0,
\end{aligned}
$$

where each of the uncertainties has 2 realizations together with their probability density mass functions. The details are

$$
\hat{a}_{11} = \begin{cases} 1, \ Pr(\{1\}) = \frac{1}{6} \\ 2, \ Pr(\{2\}) = \frac{5}{6}, \end{cases} \qquad
\hat{a}_{22} = \begin{cases} 2, \ Pr(\{2\}) = \frac{1}{4} \\ 3, \ Pr(\{3\}) = \frac{3}{4}, \end{cases}
$$

$$
\hat{b}_1 = \begin{cases} 170, \ Pr(\{170\}) = \frac{1}{3} \\ 180, \ Pr(\{180\}) = \frac{2}{3}, \end{cases} \qquad
\hat{b}_2 = \begin{cases} 160, \ Pr(\{160\}) = \frac{1}{2} \\ 162, \ Pr(\{162\}) = \frac{1}{2}, \end{cases}
$$

with penalty prices $s_1 = 5$ and $s_2 = 7$.

Exercise 214 Suppose that uncertainties $\hat{a}_{11}$ and $\hat{b}_1$ in the first constraint of the previous problem do not have probability interpretations. Instead, two realizations, 1 and 2 of $\hat{a}_{11}$ and other two realizations, 170 and 180 of $\hat{b}_1$ have random set information as follows: $m(\{1\}) = \frac{1}{2}, m(\{1, 2\}) = \frac{1}{2}, m(\{170\}) = \frac{1}{3}$ and $m(\{180\}) = \frac{2}{3}$. Generate pessimistic and optimistic stochastic expected recourse models.

Exercise 215 Prove that for any $A \subset B$, $\mathrm{Bel}(A) \leq \mathrm{Bel}(B)$. Provide an example when the equality sign happens.

Exercise 216 Provide an example that an optimal solution to a minimax regret problem is not one of the optimal solutions of realizations.

Exercise 217 Given $U = \{u_1, u_2, u_3\}$, where $pr(u_1) \in [1/3, 1/2]$, $pr(u_2) \in [1/6, 2/3]$, $pr(u_3) \in [1/4, 1/2]$, verify that the interval range of probability represents belief and plausibility or not. Moreover if $u_1 = 1$, $u_2 = 2$, $u_3 = 3$, find the minimum and maximum expected values.

Exercise 218 Generate the corresponding stochastic expected recourse model and minimax regret model (when not considering probabilities) for the following linear programming with uncertainty.

$$
\begin{aligned}
\min \quad & 2x_1 \quad +3x_2 \\
\text{s.t.} \quad & \hat{a}_{11}x_1 \quad +6x_2 \geq \hat{b}_1 \\
& 3x_1 +\hat{a}_{22}x_2 \geq \hat{b}_2 \\
& x_1, \qquad x_2 \geq 0,
\end{aligned}
$$

where each of the uncertainties has 2 realizations together with their probability density mass functions. The details are

$$\hat{a}_{11} = \begin{cases} 1, & Pr(\{1\}) = \frac{1}{6} \\ 2, & Pr(\{2\}) = \frac{5}{6}, \end{cases} \qquad \hat{a}_{22} = \begin{cases} 2, & Pr(\{2\}) = \frac{1}{4} \\ 3, & Pr(\{3\}) = \frac{3}{4}, \end{cases}$$

$$\hat{b}_1 = \begin{cases} 170, & Pr(\{170\}) = \frac{1}{3} \\ 180, & Pr(\{180\}) = \frac{2}{3}, \end{cases} \qquad \hat{b}_2 = \begin{cases} 160, & Pr(\{160\}) = \frac{1}{2} \\ 162, & Pr(\{162\}) = \frac{1}{2}, \end{cases}$$

with penalty prices $s_1 = 5$ and $s_2 = 7$.

Exercise 219 Suppose that uncertainties $\hat{a}_{11}$ and $\hat{b}_1$ in the first constraint of the previous problem do not have probability interpretations. Instead, two realizations, 1 and 2 of $\hat{a}_{11}$ and other two realizations, 170 and 180 of $\hat{b}_1$ have random set information as follows: $m(\{1\}) = \frac{1}{2}, m(\{1, 2\}) = \frac{1}{2}, m(\{170\}) = \frac{1}{3}$ and $m(\{180\}) = \frac{2}{3}$. Generate pessimistic and optimistic stochastic expected recourse models.

References

1. H. Tanaka, K. Asai, Fuzzy linear programming with fuzzy numbers. Fuzzy Sets Syst. **13**, 1–10 (1984)
2. H. Tanaka, H. Ichihashi, K. Asai, Fuzzy Decision in linear programming with trapezoid fuzzy parameters, in *Management Decision Support Systems Using Fuzzy Sets and Possibility Theory*, eds. by J. Kacprzyk, R.R. Yager (Verlag TUV, Koln, 1985), pp. 146–154
3. J.J. Buckley, Possibility and necessity in optimization. Fuzzy Sets Syst. **25**(1), 1–13 (1988)
4. J.J. Buckley, Solving possibilistic linear programming problems. Fuzzy Sets Syst. **31**(3), 329–341 (1989)
5. K.D. Jamison, W.A. Lodwick, The construction of consistent possibility and necessity measures. Fuzzy Sets Syst. **132**, 1–10 (2002)
6. W.A. Lodwick, K.D. Jamison, Interval-valued probability in the analysis of problems that contain a mixture of fuzzy, possibilistic and interval uncertainty, in *2006 Conference of the North American Fuzzy Information Processing Society, June 3-6, 2006, Montréal, Canada*, eds. by K. Demirli, A. Akgunduz, paper 327137 (2006)
7. W.A. Lodwick, K.D. Jamison, Interval-valued probability in the analysis of problems containing a mixture of possibility, probabilistic, and interval uncertainty. Fuzzy Sets Syst. **159**(1), 2845–2858 (2008). Accessed 1 Nov 2008
8. P. Thipwiwatpotjana, W. Lodwick, Pessimistic, optimistic, and minimax regret approach for linear programs under uncertainty. Fuzzy Opt. Decis. Mak. **13**(2), 151–171 (2014)
9. P. Diamond, P. Kloeden, *Metric Spaces of Fuzzy Sets* (World Scientific Publishing Company, Singapore, 1994)
10. P. Diamond, P. Kloeden, Robust Kuhn-Tucker conditions and optimization under imprecision, in *Fuzzy Optimization: Recent Advances*, eds. by M. Delgado, J. Kacprzyk, J.-L. Verdegay, M.A. Vila (Physica-Verlag, Heidelberg, 1994), pp. 61–66
11. K.D. Jamison, Modeling uncertainty using probability based possibility theory with applications to optimization. Ph.D. Thesis, UCD Department of Mathematics (1998)

12. S. Saito, H. Ishii, Existence criteria for fuzzy optimization problems, in *Proceedings of the international Conference on Nonlinear Analysis and Convex Analysis, Niigata, Japan 28–31, July, 1998*, eds. by W. Takahashi, H. Tanaka (World Scientific Press, Singapore, 1998), pp. 321–325

13. M. Inuiguchi, Stochastic programming problems versus fuzzy mathematical programming problems. Jpn. J. Fuzzy Theory Syst. **4**(1), 97–109 (1992)

14. M. Inuiguchi, Necessity measure optimization in linear programming problems with fuzzy polytopes. Fuzzy Sets Syst. **158**, 1882–1891 (2007)

15. M. Inuiguchi, On possibility/fuzzy optimization, in *Foundations of Fuzzy Logic and Soft Computing: 12th International Fuzzy System Association World Congress, IFSA 2007, Cancun, Mexico, June 2007, Proceedings,* eds. by P. Melin, O. Castillo, L.T. Aguilar, J. Kacpzryk, W. Pedrycz (Springer, 2007), pp. 351–360

16. M. Inuiguchi, Robust optimization by means of fuzzy linear programming, in *Managing Safety of Heterogeneous Systems: Decisions under Uncertainties and Risks*, eds. by Y. Ermoliev, M. Makowski, K. Marti (Springer-Verlag, Berlin, 2012), LNEMS 658, pp. 219–239

17. M. Inuiguchi, M. Sakawa, Y. Kume, The usefulness of possibility programming in production planning problems. Int. J. Prod. Econ. **33**, 49–52 (1994)

18. W.A. Lodwick, A generalized convex stochastic dominance algorithm. IMA J. Math. Appl. Bus. & Ind. **2**, 225–246 (1999)

19. W. Ogryczak, A. Ruszczyński, From stochastic dominance to mean-risk models: semideviations as risk measure. Eur. J. Oper. Res. **116**, 33–50 (1999)

20. Y. Chalco-Cano, W.A. Lodwick, A. Rufian-Lizana, Optimality conditions of type KKT for optimization problem with interval-valued objective function via generalized derivative. Fuzzy Opt. Decis. Mak. **12**(1), 305–322 (2013)

21. P. Diamond, Congruent classes of fuzzy sets form a Banach space. J. Math. Anal. Appl. **162**, 144–151 (1991)

22. K.D. Jamison, Possibilities as cumulative subjective probabilities and a norm on the space of congruence classes of fuzzy numbers motivated by an expected utility functional. Fuzzy Sets Syst. **111**, 331–339 (2000)

23. W.A. Lodwick, O. Jenkins, Constrained intervals and interval spaces. Soft Comput. **17**(8), 1393–1402 (2013)

24. W.A. Lodwick, D. Dubois, Interval linear systems as a necessary step in fuzzy linear systems. Fuzzy Sets Syst. **274**, 227–251 (2015)

25. M. Ehrgott, *Multicriteria Optimization* (Springer Science & Business, Berlin, 2006)

26. J. Jahn (ed.), *Vector Optimization* (Springer, Berlin, 2009)

27. C.V. Negoita, N. Sularia, On fuzzy mathematical programming and tolerances in planning. Econ. Comput. Econ. Cybern. Stud. Res. **1**, 3–15 (1976)

28. W.A. Lodwick, Analysis of structure in fuzzy linear programs. Fuzzy Sets Syst. **38**(1), 15–26 (1990)

29. A.L. Soyster, Convex programming with set-inclusive constraints and applications to inexact linear programming. Oper. Res. **21**, 1154–1157 (1973)

30. D. Dubois, Linear programming with fuzzy data, in *Analysis of Fuzzy Information, Vol. III: Applications in Engineering and Science*, ed. by J.C. Bezdek (CRC Press, Boca Raton, 1987), pp. 241–263

31. M. Inuiguchi, H. Ichihashi, Y. Kume, Relationships between modality constrained programming problems and various fuzzy mathematical programming problems. Fuzzy Sets Syst. **49**, 243–259 (1992)

32. A. Charnes, W.W. Cooper, Chance-constrained programming. Manag. Sci. **6**(1), 73–79 (1959)

33. M. Inuiguchi, W.A. Lodwick, Foundational contributions of K. Asai and H. Tanaka to fuzzy optimization. Fuzzy Sets Syst. **274**, 24–46 (2015)

34. Y. Lai, C. Hwang, *Fuzzy Mathematical Programming* (Springer, Berlin, 1992)

35. H.A. Simon, *A New Science of Management Decision* (Harper, New York, 1960)

36. H.A. Simon, *Models of Bounded Rationality* (MIT Press, Cambridge, Mass, 1997)

37. W. Strother, Continuity for multi-valued functions and some applications to topology. (doctoral dissertation) Tulane University (1952)
38. W. Strother, Fixed points, fixed sets, and m-retracts. Duke Math. J. **22**(4), 551–556 (1955)
39. W. Strother, Continuous multi-valued functions. Boletim da Sociedade de Matematica de São Paulo **10**, 87–120 (1958)
40. R.E. Moore, W. Strother, C.T. Yang, Interval integrals. Technical Report Space Div. Report LMSD703073, Lockheed Missiles and Space Co. (1960)
41. R.E. Moore, *Interval Analysis* (Prentice-Hall, Englewood Cliffs, 1966)
42. L.A. Zadeh, Fuzzy sets. Inf. Control **8**, 338–353 (1965)
43. H. Markowitz, Portfolio selection. J. Finance **7**, 77–91 (1952)
44. M.C. Steinbach, Markowitz revisited: mean-Variance models in financial portfolio analysis. SIAM Rev. **43**, 31–85 (2001)
45. K.D. Jamison, W.A. Lodwick, Fuzzy linear programming using penalty method. Fuzzy Sets Syst. **119**, 97–110 (2001)
46. K.D. Jamison, W.A. Lodwick, Minimizing unconstrained fuzzy functions. Fuzzy Sets Syst. **103**(3), 457–467 (1999)
47. R.R. Yager, On choosing between fuzzy subsets. Kybernetes **9**, 151–154 (1980)
48. R.R. Yager, A procedure for ordering fuzzy subsets of the unit interval. Inf. Sci. **24**, 143–161 (1981)
49. W.A. Lodwick, K. Bachman, Solving large scale fuzzy possibilistic optimization problems. Fuzzy Opt. Decis. Mak. **4**(4), 257–278 (2005)
50. S. Rivaz, M.A. Yaghoobi, Minimax regret solution to multiobjective linear programming problems with interval objective functions coefficients. Cent. Eur. J. Oper. Res. **21**(3), 625–649 (2013)
51. M. Inuiguchi, M. Sakawa, Minimax regret solution to linear programming problems with an interval objective function. Eur. J. Oper. Res. **86**(3), 526–536 (1995)
52. S. Chanas, D. Kuchta, Multiobjective programming in optimization of interval objective functions a generalized approach. Eur. J. Oper. Res. **94**(3), 594–598 (1996)
53. S. Giove, S. Funari, C. Nardelli, An interval portfolio selection problem based on regret function. Eur. J. Oper. Res. **170**(1), 253–264 (2006)
54. K. Kuhn, et al., Bi-objective robust optimisation. Eur. J. Oper. Res. **252**(2), 418–431 (2016)
55. J.R. Birge, F. Louveaux, *Introduction to Stochastic Programming* (Springer, Berlin, 1997)
56. P. Thipwiwatpotjana, W.A. Lodwick, Pessimistic, optimistic, and minimax regret approaches for linear programs under uncertainty. Fuzzy Opt. Decis. Mak. **13**(2), 151–171 (2014)
57. C.-L. Hwang, A.S. Masud, *Multiple Objective Decision Making Methods and Applications: A State-of-the-Art Survey*, vol. 164 (Springer Science & Business Media, Berlin, 2012)
58. H. Nguyen, *An Introduction to Random Sets* (Chapman & Hall/CRC, Boca Raton, 2006)
59. A.P. Dempster, Upper and lower probability induced by a multivalued mapping. Ann. Math. Stat. **38**, 325–339 (1967)
60. G. Shafer, *A Mathematical Theory of Evidence* (Princeton University Press, Princeton, 1976)
61. G. Shafer, Belief functions and possibility measures (Chap. 3), in *Mathematics and Logic: Analysis of Fuzzy Information*, vol. 1, ed. by J.C. Bezdek (CRC Press Inc, Boca Raton, 1987), pp. 51–84
62. P. Thipwiwatpotjana, W.A. Lodwick, A relationship between probability interval and random sets and its application to linear optimization with uncertainties. Fuzzy Sets Syst. **231**, 45–57 (2013)
63. E. Untiedt, Fuzzy and possibilistic programming techniques in the radiation therapy problem: an implementation-bases analysis. Masters Thesis, University of Colorado Denver (2006)